A C S   S Y M P O S I U M   S E R I E S **592**

Parallel Computing in Computational Chemistry

Timothy G. Mattson, EDITOR

Intel Corporation

Developed from a symposium sponsored
by the Division of Computers in Chemistry
at the 207th National Meeting
of the American Chemical Society,
San Diego, California,
March 13–17, 1994

American Chemical Society, Washington, DC 1995

Library of Congress Cataloging-in-Publication Data

Parallel computing in computational chemistry / Timothy G. Mattson, editor

p.   cm.—(ACS symposium series; 592)

"Developed from a symposium sponsored by the Division of Computers in Chemistry at the 207th National Meeting of the American Chemical Society, San Diego, California, March 13–17, 1994."

Includes bibliographical references and indexes.

ISBN 0–8412–3166–4

1. Chemistry—Data processing. 2. Parallel processing (Electronic computers) I. Mattson, Timothy G., 1958–   . II. American Chemical Society. Division of Computers in Chemistry. III. American Chemical Society. Meeting (207th: 1994: San Diego, Calif.) IV. Series.

QD39.3.E46P32   1995
542′.85′435—dc20   95–1232
                                                            CIP

This book is printed on acid-free, recycled paper.

PRINTED IN THE UNITED STATES OF AMERICA

MUST
IAHU 1310

Foreword

THE ACS SYMPOSIUM SERIES was first published in 1974 to provide a mechanism for publishing symposia quickly in book form. The purpose of this series is to publish comprehensive books developed from symposia, which are usually "snapshots in time" of the current research being done on a topic, plus some review material on the topic. For this reason, it is necessary that the papers be published as quickly as possible.

Before a symposium-based book is put under contract, the proposed table of contents is reviewed for appropriateness to the topic and for comprehensiveness of the collection. Some papers are excluded at this point, and others are added to round out the scope of the volume. In addition, a draft of each paper is peer-reviewed prior to final acceptance or rejection. This anonymous review process is supervised by the organizer(s) of the symposium, who become the editor(s) of the book. The authors then revise their papers according to the recommendations of both the reviewers and the editors, prepare camera-ready copy, and submit the final papers to the editors, who check that all necessary revisions have been made.

As a rule, only original research papers and original review papers are included in the volumes. Verbatim reproductions of previously published papers are not accepted.

M. Joan Comstock
Series Editor

Contents

INDEXES

Preface

A PARALLEL COMPUTER IS A SUPERCOMPUTER built from simpler computers. This includes everything from 16 heads of Cray C-90 computers to 3600 microprocessors in Intel's Paragon computers to 16,384 processing elements in a MasPar MP-2. In every case, the motivation is the same: How do you get more computing done in less time and for less money?

Getting more of anything for less money sounds like a smoke-and-mirrors trick. With parallel computing, however, it isn't a trick. Parallel computers really do provide the ultimate performance, and because they are built from simpler (and usually standard) components, they really do cost less. So why isn't all supercomputing done on parallel computers?

The answer is software. To take advantage of a parallel computer, a user needs to have software that runs in parallel. Without a "critical mass" of application software, parallel computers are computer science research machines or specialized tools for big-budget projects.

So where are we with regard to software? Serious parallel application development has been going on for a little more than 10 years. We have learned a lot in the past decade and now understand how to write software for parallel computers. But have we reached a critical mass of computational chemistry applications?

I believe we reached critical mass for parallel chemistry software within the past year. Computational chemists can now find software for most types of chemical computation. This fact is not widely known outside of a small group of parallel computational chemists, so I worked with Michel Dupuis and Steven Chin (both of IBM) to reach out to computational chemists at large with a symposium titled "Parallel Computing in Computational Chemistry."

The symposium was organized around two types of papers called *"show-and-tell"* and *"nuts-and-bolts"*. The show-and-tell papers were aimed at software users and stressed the scientific problems solved on parallel computers. The nuts-and-bolts papers were directed at software developers and covered the algorithms and software techniques used for parallel computing.

This book is based on that symposium. The connection is a loose one, however, because the book emphasizes the nuts-and-bolts papers, whereas the symposium had a more even mix of the two types of papers.

This nuts-and-bolts emphasis was not by design, but is a reflection of what the authors themselves felt to be most interesting in their work.

My hope for this book is to reach both novice and experienced parallel computational chemists. For the novice, Chapter 1 introduces the concepts and jargon of parallel computing. To round out this introduction, novice parallel programmers should also read Chapters 2, 3, 9, and 10. Chapters 2 and 3 provide an overview of the parallel ab initio program GAMESS and describe some of the large calculations it has made possible. Chapter 10 is a good introduction to how molecular dynamics codes can be simply parallelized. The chapter includes plenty of pseudocode to make the discussion as clear as possible. Finally, Chapter 9 is an excellent description of the various algorithms used in parallel molecular dynamics.

For the experienced parallel computational chemist, this book is packed with valuable information. Chapters discuss the latest trends in parallel programming tools, such as object-oriented programming (Chapters 4 and 6), tool command language (tcl) (Chapter 7), and the global arrays (GA) package (Chapter 6). Other chapters include some of the latest algorithms, such as the parallel fast multipole approximation (Chapter 11), the force decomposition algorithm (Chapter 9), and the use of distributed shared memory in post-Hartree–Fock calculations (Chapter 6).

The chapters in this book give a good feel for the range of hardware used in parallel computational chemistry: from massively parallel single instruction–multiple data (SIMD) machines (Chapter 13) to cost-effective workstation clusters (Chapter 5). They also provide a well-rounded view of what it is like to work with parallel systems—including some of the frustrations (Chapter 8).

In all, this book is a self-contained introduction to the state of the art in parallel computational chemistry. I can't claim that every important method is in here (notable omissions are Monte Carlo and density functional methods), but the most common parallel computational chemistry methods are here.

Timothy G. Mattson
Intel Corporation
Supercomputer Systems Division
Mail Stop C06-09
14924 Northwest Greenbrier Parkway
Beaverton, OR 97006

November 15, 1994

Chapter 1

Parallel Computing

Timothy G. Mattson

Intel Corporation, Supercomputer Systems Division, Mail Stop C06-09, 14924 Northwest Greenbrier Parkway, Beaverton, OR 97006

Computational chemists place tremendous demands on their computers. From the thousands of atoms in molecular modeling to the hundreds of basis functions in quantum chemistry, chemists are among the most demanding of all supercomputer users. It is therefore not surprising that computational chemists consistently find themselves at the forefront of high performance computing.

Over the last decade, the *forefront of high performance computing* has come to mean parallel computing: i.e. the use of many processors in unison to solve a single problem. These *parallel computers* not only provide the most computational power, they are also more cost-effective than traditional vector-based supercomputers. More cost effective, that is, in terms of hardware. When software costs are factored in, a different picture emerges.

The software costs for parallel systems are excessive. The reason for this is simple: parallel *programming is hard*! On a traditional supercomputer, a compiler can look at C or Fortran code and find operations to compute on the system's vector units. Thus, these computers can be used with minimal additional programming. On the other hand, parallel computers need software that has been decomposed into relatively independent tasks. This decomposition is heavily dependent on an algorithm's structure and so complex that it is unlikely compilers will ever be able to do the job automatically. Therefore, to use parallel computers, one must write parallel programs.

Even though the programming costs are great, computational chemistry applications have been slowly but surely moving onto parallel architectures. After many years of difficult programming, chemists can now find parallel software for just about every class of chemical computation.

To understand these developments and perhaps get involved in bringing even more applications to parallel computers, it is worthwhile to step back and take a close look at parallel computing. That is the goal of this chapter. We will explore parallel architectures, parallel algorithms, and the ways parallel programmers evaluate the performance of parallel algorithms. Throughout the discussion, careful attention will be paid to the jargon of parallel computing so the reader can jump directly from this chapter into the literature of parallel computing.

0097–6156/95/0592–0001$12.00/0

What is Parallel Computing?

The term *parallel computing* refers to computing that uses multiple central processing units (CPU) to solve a single problem. The hardware that supports parallel computing goes under a number of names: multicomputer, parallel computer, cluster, multiprocessor, etc. Each of these names suggests a particular nuance of architecture. We won't worry about these details, however, and will use the term *parallel computer* to mean any system with multiple CPU's. We will refer to the individual processing units as the *nodes* of the parallel computer.

There are many different ways to combine CPU's into a parallel computer. To keep track of these options, computer scientists organize parallel architectures in terms of instruction streams and data streams (1). Two cases have become everyday terms to the parallel programmer:
1. Single Instruction, Multiple-Data (SIMD).
2. Multiple-Instruction, Multiple Data (MIMD).

A SIMD computer consists of multiple nodes working in lock-step from a single instruction stream. While this accurately describes some super-scalar and vector architectures, parallel programmers reserve the term SIMD for computers containing a very large number (thousands to tens of thousands) of simple processors with their own local memory. Since all of the processors are driven by a single instruction stream, the parallelism is expressed in terms of concurrent operations on distinct data elements.

SIMD computing's single instruction stream makes the programmer's job easier; a belief that has driven much of the interest in the SIMD architecture. Using a single instruction stream, however, carries a price. Whenever a SIMD program contains conditional logic, some nodes execute while others remain idle. For example, depending on the data, the run-time for an IF-ELSE structure can be equal to the sum of the run-times for the individual IF and ELSE clauses. Hence, while the SIMD program may be easy to write, getting the most out of the computer may require complicated coding to eliminate conditional logic.

The best example of a SIMD computer is the MP-2 computer from MasPar Computer Corporation. This machine has from 1024 to 16,384 32 bit processors each with its own memory. All of the processors work off a single instruction stream provided by a single array control unit. The MP-2 is tightly coupled to a front-end workstation. The programs are written in a data parallel dialect of a sequential language that has a rich set of array based operations (such as Fortran90) with sequential operations occurring on the front-end workstation and array operations occurring on the MP-2. If the programmer is not careful, *data sloshing* occurs meaning that excess data movement occurs between the MP-2 and the front-end workstation.

MIMD computers are based on a more general parallel architecture with processing elements that have their own instruction and data streams. In most cases, a MIMD computer is built with microprocessor components developed for the PC and workstation markets. The huge volume of these markets fuels fiercely competitive R&D efforts that keep these standard components at the leading edge of performance. These same market forces keep component prices low allowing MIMD computers to easily hold the price performance lead in supercomputing.

MIMD systems are divided into two categories: *shared memory* and *distributed memory*. Good examples of shared memory MIMD computers are the systems marketed by Silicon Graphics Incorporated (SGI). These computers have multiple processors connected to a shared memory by a high speed bus. This gives the programmer a single address-space which simplifies programming because data is where it is needed when it is needed. On the other hand, a single address space complicates programming since processes can corrupt each other's data. To solve this problem, programmers place semaphores around critical data elements to enforce a safe order for memory references.

Shared memory computers are usually limited to peak sizes of 16 to 32 nodes. This limitation exists because the bus connecting the processors to the shared memory saturates if too many nodes are added. To build computers with large numbers of nodes (a so called scalable system), the aggregate access rate to memory must increase as more nodes are added. Distributed memory computers provide a solution to this problem.

As the name implies, distributed memory MIMD computers locate the memory with each processor. They may provide a shared memory programming model (T3D from Cray Research Inc.), but the underlying architecture uses multiple independent processors with their own local memory. These processors are connected by a communication network that supports passing messages between individual nodes (hence why these sometimes are called *message passing* architectures). The network connects the processors in a particular arrangement such as a mesh (Paragon and Delta Supercomputers from Intel Corporation), the fat tree (CM-5 from Thinking Machines Corp.) a hypercube (nCUBE 2 from nCUBE Corporation, iPSC/860 from Intel Corporations) or a hierarchy of crossbar switches (SP1 and SP2 from IBM). This arrangement of processors is called the computer's *topology*.

Programmers used to pay a great deal of attention to a parallel computer's topology. Fortunately, most distributed memory MIMD systems now use sophisticated message routing mechanisms that let a processor communicate to any other processor on the same time scale. This is, of course, only an approximation and for performance tuning the detailed arrangement of processors can still be important. These are low-level optimizations, however, so most parallel programmers can safely ignore a computer's topology.

A sub-class of distributed memory MIMD computers is *workstation clusters*. As the name implies, these parallel computers are built from a network of workstations. In many cases, ethernet-connected networks of desktop workstations can be used for impressive supercomputing off-hours when they are otherwise idle (see chapter 5). Workstation clusters can also be built specifically for parallel computing. In this case, sophisticated interconnection hardware based on fiber-optic networks (FDDI) or a high speed switch such as an ATM switch can be used (see chapter 11). The interest in this type of parallel computing has grown dramatically in the last five years as software (2,3) has become available to make these systems as easy to program as traditional parallel computers.

The research described in this book was carried out on a wide range of parallel systems. Table I lists these systems along with the chapters where they appear. This table shows a predominance of systems from Intel Corp. This is due to the popularity of the distributed memory MIMD architecture and the fact that Intel has been building

Table I. Parallel computers discussed in this book. The vendors for these systems are named in the text.

Computer	nCUBE2	T3D	C90	CM5	MP2	Delta	iPSC/860	Paragon	SGI	Clusters
Architecture	DM MIMD	DM, DSM, MIMD	SM, MIMD	DM, MIMD	DM, SIMD	DM, MIMD	DM, MIMD	DM, MIMD	SM, MIMD	DM, MIMD
Chapters	9	9	14	14	13	3, 6, 12	3, 6, 8, 9	3, 4, 8, 9, 11	7	2, 5, 10, 11

DM   Distributed Memory
SM   Shared Memory
DSM   Distriubted Shared Memory
MIMD Multiple Instruction, Multiple Stream Architecture
SIMD Single Instruction, Multiple Data Architecture

this type of computer longer than other vendors (with the possible exception of nCUBE Corp.). The second most common system in this book is workstation clusters. Clusters do not provide the ultimate performance, but they are ubiquitous and a good source of affordable supercomputing.

For many years, a *SIMD-vs.-MIMD* debate raged within the parallel computing community. There are still echoes of this debate, but essentially its over and the MIMD-camp won. This can be seen by the lack of SIMD systems in Table I, but more importantly by the poor sales of SIMD computers in the marketplace. These systems were supposed to be easy to program, but it turned out that optimizing SIMD programs was very difficult. At the time this is being written, only one manufacturer continues to produce general purpose supercomputers based on the SIMD architecture, making MIMD systems the overwhelming majority. Hence, while the vocabulary and general concepts discussed in the rest of this chapter apply to both architectures, the bulk of this discussion is specialized to MIMD computers. To learn more about the use of SIMD computers, see chapter 13.

How to Program Parallel Computers

Writing software is a complex undertaking regardless of the target system. If that target computer is parallel, however, it quickly becomes a truly daunting task. This difficulty has hindered the adoption of parallel computing for mainstream supercomputing.

To help understand parallel programming, lets first take a look at programming in general. Programmers view a computer in terms of a high level abstraction called a *programming model*. This frees them from low-level, system dependent details and lets them write portable software. For single processor computers, there is a common programming model that virtually all programmers use: the *von Neumann model*. The von Neumann model views a computer as a single processor with a single stream of instructions that operates on a single memory. Processors and the memory sub-systems vary widely from one computer to another. These details can be neglected by the programmer (except for final performance tuning), letting a program coded to the von Neumann model run on any single processor computer.

Unfortunately, parallel programmers have not converged on a single programming model. This complicates a programmer's job forcing him or her to choose from a range of programming models. In addition, the lack of a universal programming model has diluted the efforts of programming tool developers resulting in relatively immature tools for parallel computing. The result is that the parallel programmer's difficult job is made even harder.

Of the many programming models for MIMD computers, most fall into one of two camps:

    1.   Task parallelism
    2.   Data Parallelism

In *task parallelism*, parallelism is expressed by mapping different actions onto different nodes. For example, the Strand parallel programming language (4) supports concurrency by executing different routines on different nodes. Another example of task parallelism is pipeline algorithms (see chapter 7). These algorithms consist of an ordered set of tasks (called stages) that execute in parallel. Input data enters the pipeline at one end, and after working through each of the stages, the final result comes out the other end. Once the pipeline is full, the algorithm proceeds with concurrency equal to the number of stages (the depth of the pipeline). In both examples, it is the tasks to be executed in parallel that guides the algorithm's design.

Algorithms using *data parallelism* are designed in terms of the data and how it is distributed among the nodes. At one extreme are pure data parallel, SIMD programs where every node applies the same stream of instructions to its own data. The data parallel model is far more general than the SIMD case, however, and includes programs with very different actions occurring on each node.

The most common data parallel programming model is called SPMD or *Single Program Multiple Data*. In this case, the same program is loaded onto each of the parallel computer's nodes. This simplification helps the programmer tremendously because only a single program needs to be written. Unlike the SIMD case, however, different operations execute concurrently from one node to another due to conditional logic within the program.

Programming models are important – especially to computer scientists trying to find more humane ways to program parallel computers. Application programmers, however, are more concerned with the implementation of a programming model; i.e. the *programming environment*. Since the data parallel model dominates parallel computing, programming environments supporting the data parallel model are by far the most common. These programming environments take a number of forms, but most share a basic structure. We refer to this basic structure as the *coordination model*. In the coordination model, a parallel program is seen as a number of sequential processes with their own local memories that coordinate their actions at specific points in the program.

For example, *coordination libraries* such as PVM (5), TCGMSG (6), or MPI (7) use the exchange of discrete messages to coordinate processes. The bulk of the program is traditional C or Fortran with library function calls to exchange messages, synchronize processes, or to spawn new processes. Because of their focus on message passing, these systems are frequently called message passing libraries. The term is too restrictive, however, since these systems do far more than exchange messages.

A more sophisticated approach uses compiler support for coordination through a *coordination language*. Coordination languages separate computation (which remains in the domain of the sequential language) from parallelism (which remains strictly within the coordination language). There are several advantages to getting the compiler involved with coordination. The compiler can detect inconsistencies in the coordination operations making initial debugging much easier. In addition, the coordination language's high level view of the parallel computer provides additional algorithmic flexibility. For example, the best known coordination language is Linda (8). In Linda, coordination takes place through a small set of operations that manipulate objects within a distinct shared memory (for more information about Linda, see chapters 5 and 10). The shared memory supports algorithms that use high level constructs such as distributed data structures and anonymous communication (i.e. the sender and/or receiver don't know the identity of one another). Linda isn't the only coordination language. Several others are available including Fortran-M (9) and PFortran (10).

While less commonly used, programming environments that do not use the coordination model are available. These programming environments are based on formal models of the parallel computer resulting in inherently parallel programming languages. For example, there are parallel programming environments that use concurrent logic programming (Strand (4)), functional programming (SISAL (11)), and SIMD-style data parallel programming (HPF (12)). Each of these systems are based on formal models of the parallel computer and have distinct advantages. To use these environments, however, a programmer must learn a new language. Programmers are reluctant to learn new languages, so these inherently parallel languages have seen insignificant usage compared to programming environments based on the coordination model.

Which programming environment is the best? This question has been addressed for some systems (13,14), but in the final analysis, a general answer does not exist. Every programmer must choose for themselves based on the sorts of algorithms that will be implemented and the parallel systems that the software will run on.

Algorithms for Parallel Computing

The heart of any program is its algorithms. The parallel programmer must deal with two levels of algorithms. First, each node runs a local program, so all the challenges of sequential algorithms must be faced. Second, there are a myriad of issues unique to parallel computing such as balancing the work load among all the nodes and making sure that data is where it is needed when it is needed. Because of these two levels of complexity, parallel algorithms are among the most challenging of all algorithms. Complicating matters further, many different parallel algorithms are available. It is easy for parallel programmers - both novice and expert - to become overwhelmed.

Fortunately, most parallel algorithms can be more easily understood by mapping them into one or more of three simple algorithm classes. We will call these algorithm classes and the code constructs that implement them, *algorithmic motifs*. The three most common algorithmic motifs are:

1.   Loop Splitting
2.   Domain Decomposition
3.   Master/worker or the task queue

In addition to the algorithmic motif, the parallel programmer must understand an algorithm's granularity. *Granularity* refers to the ratio of the time spent computing to the time spent communicating (or synchronizing). If an algorithm must communicate after a small amount of computation, it is called *fine grained*. If a great deal of computation occurs for each communication, the algorithm is said to be *coarse grained*.

Granularity is also used to describe the number of simultaneous processes within a program. If an algorithm can use only a small number of simultaneous processes, the program is called coarse grained: even if it requires a great amount of communication relative to computation. Usually a program is coarse or fine grained under both definitions of granularity, but this isn't always the case.

It is important to understand the granularity of an algorithm and make sure it is consistent with the granularity of the hardware. For example, if the hardware communication rate is much slower than the computation rate (such as an ethernet connected workstation cluster), then fine grained algorithms will not run well. Of course, communication capacity can be under utilized so coarse grained algorithms work well on fine grained parallel computers.

Notice that it is the granularity, not the amount of communication, that governs the effectiveness of an algorithm on a particular parallel computer. For example, some parallel programmers assume that a collection of workstations on a local area network can not be used with algorithms that require significant communication. This isn't true! If computation grows faster than communication as a problem's size increases, then it is possible to increase a problem's size so its granularity matches the coarse granularity of a workstation cluster. Such large problem sizes may not be interesting, but when they are, it is possible to do supercomputing on a workstation cluster - even if substantial communication is required. Hence, it isn't the amount of communication but the ratio of computation to communication (granularity) that matters.

We will now look at each of these algorithmic motifs in detail. For each case, we will describe what the motif is, when it can be used, and finally, how it is used to code a parallel program.

Loop Splitting. The parallelism in a loop splitting algorithm comes from assigning loop iterations to different processors. It is almost always used within a *replicated data SPMD program*. This means that the same program is loaded onto each node of the parallel computer and that key data structures are replicated on each node. At the conclusion of the split loops, a single copy of this data is rebuilt on each node. It is this reconstruction that represents the communication phase of the parallel algorithm. Data replication is a powerful technique and is a simple way to assure that the right data is located where it is needed when it is needed.

The loop splitting algorithm can be used whenever:

1. The bulk of a program's run time is spent in a few loops.
2. The iterations of the split loops are independent and can execute in any order.
3. The replicated data fits in each node's memory.
4. The amount of data that must be replicated is small enough so communication doesn't overwhelm computation.

A simple example will clarify the loop splitting algorithmic motif and show how it is used. Consider the following code fragment:

```
do i = 0, NUMBER_OF_ITERATIONS
    call WORK()
end do
```

If the operations carried out within WORK() are independent of any previous loop iterations (i.e. there are no loop carried dependencies) this code can be parallelized with loop splitting. First, the same program is loaded onto each node of the parallel computer (the SPMD program structure). Next, logic is added to replicate any key data structures manipulated in this loop. The loop iterations are then spread out among the nodes in some manner. A common trick is to use a *cyclic distribution* of the loop iterations:

```
do I = ID, NUMBER_OF_ITERATIONS, NUM_NODES
    call WORK()
end do
call GLOBAL_COMBINE()
```

where we assume that each of the NUM_NODES processors has a unique node ID ranging from 0 to NUM_NODES-1. The cyclic distribution assigns loop iterations as if a deck of cards were being dealt to the nodes with each node getting iterations ID, ID+NUM_NODES, ID+2*NUM_NODES, etc. As the calculation proceeds on each node, it fills in a scattered subset of any replicated data structures. When the loop is finished on each node, this scattered data is recombined into a globally consistent data structure with a call to a GLOBAL_COMBINE() operation. This operation uses *all-to-all communication,* i.e. each node contributes its subset of the data to each of the other nodes. Since all nodes must participate, GLOBAL_COMBINE() operations implicitly invoke a *synchronization barrier* - i.e. a point in a parallel program where each node waits until all nodes have arrived.

All of the communication in the loop splitting algorithm occurs in the GLOBAL_COMBINE operation. Of the many GLOBAL_COMBINE() operations, the most common is the *global summation.* The starting point for a global summation is distinct (though same sized) vectors on each node. The corresponding elements of the vector are summed together leading to a single vector containing the summed elements. The operation concludes by replicating the vector on each of the nodes of the parallel computer using a broadcast or in the most clever algorithms, the vectors are manipulated so the same reduced vector is produced in parallel on each of the nodes (15,18). While it is easy to describe a global combine operation, writing one that works efficiently and correctly is difficult. Fortunately, these operations are included in most parallel programming environments. For more information about global summations including code for a primitive method, see chapter 10.

The cyclic distribution is not the only way to assign loop iterations. On some architectures, reuse of data from the cache is maximized by having a *blocked distribution* with contiguous blocks of loop indices assigned to each node. One way to code this is to use arrays indexed by the node ID to indicate the first and last loop indices for each node. For example:

```
do I = FIRST(ID), LAST(ID)
      call WORK()
end do
call GLOBAL_COMBINE()
```

The disadvantage of the blocked distribution is its potential to produce uneven amounts of computing among the nodes. If different iterations take different amounts of time, then processors can run out of work at different times. The cyclic distribution avoids this problem in a statistical manner due to the scattering of the loop iterations among the nodes. A program that uses a blocked distribution, however, may need to periodically recompute the FIRST and LAST arrays to keep all of the processors evenly loaded (dynamic load balancing).

Any algorithm that depends on a replicated data approach suffers from communication that scales poorly and excess memory utilization. This limits the scalability of an algorithm (i.e. the number of nodes that can be effectively used in the computation). These are serious weaknesses for a parallel algorithm, yet loop splitting is by far the most common parallel algorithmic motif used by computational chemists. Why is this the case?

Loop splitting is so common for one reason: simplicity. Given a complex program that has evolved over many years (and many programmers) the loop splitting algorithm lets one create a parallel program with minimum changes to the original code. It also lets one parallelize a program without understanding how its data structures are manipulated. Eventually, as programs are written from scratch for parallel computers, loop splitting algorithms will be used less often. But for the immediate future, sequential programs must be ported to parallel platforms, and the loop splitting algorithms will continue to dominate. For examples of the loop splitting motif, see chapters 2, 8, 9, and 10 as well as the classic paper on the parallelization of CHARMM (18).

Domain Decomposition. The central organizing principle of a *domain decomposition* (or *geometric decomposition*) algorithm is the way data is broken down into smaller units (the *data decomposition*). Once this decomposition is carried out, a program operates locally on its chunk of the data. Communication occurs at the boundaries of the local domains and is usually restricted to neighboring processors. This is the inherent advantage of these methods. By eliminating global communication, domain decomposition methods can use more nodes. Furthermore, these algorithms use memory more efficiently since they only need space for a local domain - not an entire copy of the global data.

The loops in domain decomposition programs run over local indices so these programs can look like block decomposition, loop splitting programs. They are quite different, however, since the domain decomposition programs must decompose the data into local blocks and communicate to selected nodes rather than globally.

Domain decomposition algorithms can be used whenever computations are localized over well defined blocks of data. Another factor to look for when choosing a domain decomposition algorithm is that the communication required to update a local data block is restricted to a small number of nearby processors.

Good examples of the domain decomposition algorithm are spatial decomposition algorithms for molecular dynamics (see chapter 9). For these algorithms, 3D space is divided into distinct regions which are mapped onto the nodes of the parallel computer. Each node updates the forces and coordinates for atoms in its region. Communication arises from two sources. First, to compute the forces for atoms near the domain's edge, atomic data is required from the neighboring domains. Second, atoms must be sent to neighboring processors when they move across a domain boundary.

Domain decomposition algorithms are significantly more complicated than loop splitting algorithms. They are usually superior algorithms in terms of effective utilization of the parallel computer, so they should be used whenever the extra effort is justified (e.g. library routines such as parallel eigensolvers) or when a program is written from scratch for a parallel computer.

An important trend in domain decomposition algorithms is to simplify the data decomposition through *distributed shared memory*. This usually is implemented as a software package that provides a restricted form of shared memory regardless of the underlying hardware's memory organization. An important example of this type of programming can be found in chapter 6 where the GA package running on top of TCGMSG is described.

Master-worker. *Master-worker* (or *task queue*) algorithms distribute independent tasks among the nodes of a parallel computer. While the other two motifs are expressions of a data parallel programming model, master worker algorithms are examples of task parallelism.

Master-worker algorithms are useful when a program consists of a large number of completely independent tasks. These sorts of problems are officially designated as *embarrassingly parallel* (16) since the parallelism is so simple to extract. For reasons that will become clear in the following paragraphs, there are such striking advantages to the master-worker algorithmic motif, it should be used whenever possible.

Logically, a master-worker program consists of two types of processes - a *master* and a *worker*. The master process manages the computation by:

    1.   Setting up the computation.
    2.   Creating and managing a collection of tasks (the *task queue*).
    3.   Consuming results.

The worker process contains some type of infinite loop within which it:

    1.   Grabs a task and tests for termination.
    2.   Carries out the indicated computation.
    3.   Returns the result to the master.

Termination is indicated in a number of ways. One approach is for the master or some worker to detect the last task and then create a *poison pill*. The poison pill is a special task that tells all the other workers to terminate. Another approach is for each task to be sequentially numbered and for each worker to check when that number of tasks has been met (or exceeded).

There are many variations of the basic master-worker motif. If consuming results is trivial or easily delayed to the end of the computation, it is quite simple to modify the master to turn into a worker after it sets up the task queue. In another variation, the generation of tasks can be spread among the workers. Finally, when the master is not required to do anything special with either the creation of tasks or consumption of results, it is possible to completely eliminate the master and replace it by a mechanism to manage a queue of tasks. For example, in the programming environment TCGMSG (6) a special process is provided that maintains a globally shared counter. One can then create an SPMD program which uses the global counter to maintain the task queue. An example of this technique can be found in chapters 2 and 6.

There are a number of advantages associated with master-worker algorithms. First, they are very easy to code. A worker can be simply created from an original sequential program by just adding logic to interact with the task queue. Ease of programming is an important advantage. Even without this advantage, there is a compelling reason to use this algorithmic motif when it is possible to do so. A master-worker program can be constructed such that it automatically balances the load among the nodes of the parallel computer.

Lets consider a worse case scenario. Consider a parallel computer for which each node has a different speed of computation. Furthermore, let the computational requirements of each task vary significantly and unpredictably. In this case, any static distribution of tasks is guaranteed to produce a poor load balance. A master-worker algorithm, deals quite easily with this situation. The workers grab tasks and compute them at their own pace. A faster node will naturally grab more tasks and therefore balance the load. Furthermore, nodes that happen to grab more complex tasks will take more time and access the task-queue less frequently. Once again, the number of tasks is naturally reduced for these more heavily loaded nodes.

Algorithms with these characteristics automatically provide *dynamic load balancing*. There are a couple conditions that must be met by the task queue in order for this motif to be most effective. First, the number of tasks must be greater than the number of nodes — preferably much greater. This holds because the amount of available parallelism is given by the number of tasks. Hence, once the tasks are all assigned, no further parallelism is available to the system.

The second condition for a fully optimum master-worker algorithm is for the longest tasks to be handled first. If the long tasks are not handled until late in the computation, a single process can be stuck working on a long task while no other tasks remain for the other nodes. By handling the long tasks first, the odds are greatest that work will be available for the other nodes during computation on the long tasks.

Master-worker algorithms are not without their shortcomings. As mentioned earlier, they really only map cleanly onto embarrassingly parallel problems. More fundamentally, the master-worker algorithm ignores the underlying system topology. While it is good to de-emphasize topology when first writing a parallel program, it can be vital to include topology during final code optimization. In some cases significant performance benefits can result by controlling which tasks are mapped onto which nodes – a level of control that master-worker algorithms do not easily permit.

Even with these shortcoming, however, the master-worker algorithm is extremely useful. Computational chemists are quite fortunate that many important algorithms can be mapped onto the master worker algorithmic motif. Most problems involving stochastic optimization (e.g. DGEOM (17)) can be mapped onto this algorithmic motif.

How is Performance Measured?

Parallel computers are used to achieve greater performance, so any discussion of parallel computing eventually must address the performance of the system.

There are several standard measures of a parallel algorithm's performance. Before describing these, consider the characteristics of a parallel application that lead to high performance. To most effectively extract performance from a parallel computer, the computational work (or load) must be evenly distributed about the nodes of the parallel computer. We use the term *load balance* to describe this situation. Algorithms with poor load balancing result in computations where some nodes are busy while others remain idle. *Static load balancing* is used when the load is computed once and remains fixed as the calculation proceeds. *Dynamic load balancing* occurs when the load is changed in the course of the calculation to keep all nodes equally occupied.

Even when the load is perfectly balanced, the performance of a parallel program will be poor if too much time is spent communicating rather than doing useful computation. This is an important effect that plays a key role in limiting how many nodes can be used. To see this point, consider the distribution of a fixed amount of work among the nodes of a parallel computer. As more nodes are used, less work is available for each node. As more nodes are added, however, communication usually remains either fixed or in some cases increases. Eventually, more time is spent communicating than computing and the performance suffers.

With these effects in mind, we can look at how performance of a parallel computer is measured. The most fundamental measurement is *speedup*. Speedup is the multiplier indicating how many times faster the parallel program is than the sequential program. For example, if the program took T_{seq} seconds on one node and $T(N)$ seconds on N nodes, the speedup is the ratio:

$$S = \frac{T_{seq}}{T(N)}$$

When the speedup equals the number of nodes in the parallel computer, the speedup is said to be *perfectly linear*.

From the speedup, we can derive an important relationship describing the maximum performance available from a parallel algorithm. This relation is called *Amdahl's law*. Amdahl's law holds because parallel algorithms almost always include work that can only take place sequentially. From this sequential fraction, Amdahl's law provides a maximum possible speedup. For example, consider the parallelization of a sequential program. If we define the following variables:

T_{seq} = time for the sequential program
α = fraction of Tseq dedicated to inherently sequential operations
γ = fraction of Tseq dedicated to parallel operations
S_{max} = maximum possible speedup
P = Number of nodes

the best possible speedup for any number of processors is:

$$S = \frac{T_{seq}}{\alpha\, T_{seq} + \dfrac{T_{seq}\, \gamma}{P}} = \frac{1}{\alpha + \dfrac{1-\alpha}{P}}$$

In the limit of infinite number of processors, this expression becomes:

$$S_{max} = \frac{1}{\alpha}$$

This is a serious constraint and was used for years to argue against parallel processing. If the sequential fraction is 10%, the best possible speedup is 10. Even a rather extreme case of a 99% parallel program gives a best possible speedup of only 100.

Amdahl's law is real and must always be considered when trying to evaluate the quality of a parallel program. However, this pessimistic view misses one key point. As the number of available processors grows, the size of the problem can grow as well. In other words, parallel computers provide speed, but they also provide the memory capacity to support larger problems.

Another way to describe the performance of a parallel program is the *efficiency*. Qualitatively, efficiency measures how effectively the resources of the multiprocessor system are utilized. Quantitative definitions of efficiency generally take the form:

$$\varepsilon = \frac{t_{ref}}{P\, t_{par}}$$

where P is the number of nodes, t_{ref} is some sequential reference time, and t_{par} the parallel time. The most rigorous definition of efficiency sets t_{ref} to the execution time for the best sequential algorithm corresponding to the parallel algorithm under study. When analyzing parallel programs, "best" sequential algorithms are not always available, and it is common to use the runtime for the parallel program on a single node as the reference time. This can inflate the efficiency since managing the parallel computation always (even when executing on one node) incurs some overhead.

Conclusion

Parallel programming is a complex art. The parallel programmer must deal with all of the problems of sequential programming, as well as a host of new problems unique to parallel computing. These uniquely parallel problems are complex and can be very difficult to master.

Parallel computing, however, is no different than many subjects and follows an "80-20 rule". In other words, 80% of the understanding comes from 20% of the knowledge. The problem is to find that key 20%; a problem this chapter has tackled and hopefully solved.

We close this chapter by emphasizing four key simplifications for the person just entering the field of parallel computing. First, view parallel computers in terms of a spectrum of MIMD systems distinguished by the granularity of the hardware. This does omit some architectures such as SIMD computers, but these systems are becoming increasingly rare. A *MIMD spectrum* outlook helps one write more effective code by putting architecture dependent details such as topology in their place; i.e. as a final optimization and not as the key focus of a programming effort.

Second, one should pick a portable programming environment they are comfortable with and stick with it. This environment should be selected based on ease of use and effectiveness for the algorithms you are interested in. Performance differences are usually not significant among the common programming environments (14).

Third, when faced with a new parallel algorithm, try and map it into some combination of the algorithmic motifs described in this chapter:
1.   Loop Splitting.
2.   Master Worker (Task Queue).
3.   Domain Decomposition.

It is not always possible to clearly map an algorithm into one of these motifs (for example, see chapters 4 and 6), but the motifs can help organize your reasoning about the algorithm.

Finally, when thinking about a parallel program, evaluate your observed performance in terms of Amdahl's law. If the load balancing is right and the problem size is large enough, your program should follow the speedup curves given by Amdahl's law. If your performance is than that predicted by Amdahl law, the load balancing is wrong or the program's sequential fraction changes unfavorably as more nodes are included in the computation.

Even with these four simplifications, parallel computing can be overwhelming. It is worth the effort, though, since chemistry stands to gain so much from parallel computing.

Numerous trademarks appear in this chapter. In each case, these trademarks are the property of their owners.

References

1.   M.J. Flynn, "Some Computer Organizations and Their Effectiveness," *IEEE Trans. computers, vol C-21*, No. 9, 1972.

2.   L. Turcotte, "A Survey of Software Environments for Exploiting Networked Computing Resources," *Tech Report # MSM-EIRS-ERC-93-2*, Mississippi State University, 1993.

3.   D. Y. Cheng, "A Survey of Parallel Programming languages and Tools," *NASA Ames Research Center Technical Report RND-93-005*, 1993.

4.   I. Foster and S. Taylor, *Strand: New Concepts in Parallel Programming*, Prentice Hall, 1990.

5.   V. Sunderam, "PVM: a Framework for Parallel Distributed Computing," *Concurrency: Practice and Experience, vol 2*, pp. 315-339, 1990.

6.   R. J. Harrison, "Portable Tools and Applications for Parallel Computers," *Int. J. Quantum Chem, vol 40*, pp. 847-863, 1991.

8.   N. Carriero and D. Gelernter. *How to Write Parallel Programs: A First Course*, MIT press, 1991.

7.   D.W. Walker, "The Design of a Standard Message Passing Interface for Distributed Memory Concurrent Computers," *Parallel Computing, vol 20*, p. 657, 1994.

9.   I. Foster, R. Olson, and S. Tuecke, "Programming in Fortran M," *Technical Report ANL-93/26*. Argonne National laboratory, 1993.

10.   B. Bagheri, T.W. Clark and L.R. Scott, "PFortran (a parallel extension of Fortran) reference manual." *Research Report UH/MD-119*, Dept. of Mathematics, University of Houston, 1991.

11.   J.T. Feo, D.C. Camm, and R.R. Oldehoeft, "A Report on the SISAL Language Project," *Journal of Parallel and Distributed Computing, vol 12*, p. 349, 1990.

12.   The HPF Forum, "High Performance Fortran, Journal of Development," Special issue of *Scientific Programming , vol. 2*, No. 1,2, 1993.

13.   T.W. Clark, R.v. Hanxleden, K. Kennedy, C. Koelbel, and L.R. Scott, "Evaluating Parallel Languages for Molecular Dynamics Computations," *Proceedings of the Scalable High Performance Computing Conference, (SHPCC-92)*, p 98, 1992

14.   T. G. Mattson, "Programming Environments for Parallel and Distributed Computing: A Comparison of p4, PVM, Linda, and TCGMSG," *International Journal of Supercomputing Applications*, to appear in 1995..

15.   R. A. van de Geijn, "Efficient Global Combine Operations," *Proceedings Sixth Distributed Memory Computing Conference*, p. 291, IEEE Computer Society Press, 1991.

16.   G. C. Fox, "Parallel Computing comes of Age: Supercomputer Level Parallel Computations at Caltech," *Concurrency: Practice and Experience, vol. 1, No. 1*, p. 63, 1989.

17.   T.G. Mattson and R. Judsen, "pDGEOM: a Parallel Distance Geometry Program," *Portland State University, C.S. Tech Report*, 1994.

18.   R. R. Brooks and M. Hodoscek, "Parallelization of CHARMM for MIMD Machines," *Chemical Design Automation News, vol 7*, p. 16, 1992.

RECEIVED December 28, 1994

Chapter 2

Parallel Implementation of the Electronic Structure Code GAMESS

Theresa L. Windus[1], Michael W. Schmidt[2], and Mark S. Gordon[2]

[1]Department of Chemistry, Northwestern University,
Evanston, IL 60208–3113
[2]Department of Chemistry, Iowa State University, Ames, IA 50101

This paper outlines various tools and techniques for the parallelization of quantum chemistry codes; in particular, for the electronic structure code GAMESS. A general overview of the parallel capabilities of GAMESS are also presented.

The parallelization of quantum chemistry codes has become a very active area of research over the last decade(1,2,3,4). Until recently, most of this research has dealt with self-consistent field (SCF) theory(1). However, in the last few years parallel implementations of post-SCF methods have been presented (2). Most of the post-SCF methods and analytic Hessians for SCF wavefunctions face the substantial problem of parallelizing the atomic orbital (AO) integral to molecular orbital (MO) integral transformation (3).

The objective of this paper is to provide general information about the parallel implementation of GAMESS. The following sections are presented in this paper: (A) a brief overview of the functionality of the *ab initio* code GAMESS (General Atomic and Molecular Electronic Structure System); (B) a short discussion of the model, software, and general ideas used to parallelize GAMESS; (C) specifics concerning the parallelization of the SCF; (D) discussion concerning the AO to MO integral transformation; (E) the transformation as applied to multi-configuration SCF (MCSCF); (F) the transformation as applied to analytic Hessians; (G) a brief overview of the parallel MP2 code; and (H) conclusions and future areas of research will be discussed.

A. Brief overview of GAMESS

GAMESS is a general electronic structure code for the determination of energies, stationary states, frequencies and various other atomic and molecular properties(1a). The wavefunctions that are available in GAMESS are given in Table I, together with information about the availability of analytic determination of gradients, hessians

0097–6156/95/0592–0016$12.00/0

(energy second derivatives with respect to the nuclear coordinates), second order Moller-Plesset theory (MP2)(*5*), and configuration interaction (CI)(*6*).

Where analytical gradients are available, GAMESS can be used to calculate stationary points (structural minima and maxima), intrinsic reaction coordinates (IRCs) between transition states and minima, and numerical Hessians. Complete details of GAMESS can be found in reference *1a*.

Table I: Tabulated overview of GAMESS

	Energy	Gradient(a)	Hessian(a)	MP2	CI	Semi(b)
RHF(c)	X	X	X	X	X	X
UHF(d)	X	X		X		X
ROHF(e)	X	X	X	X	X	X
GVB(f)	X	X	X		X	
MCSCF(g)	X	X			X	

a. Refers to analytic evaluation. Numerical Hessians are available whenever analytic gradients are available.
b. Semi-empirical wavefunctions: AM1, MNDO, PM3(*7*). Energies and analytic gradients are available.
c. Restricted Hartree-Fock, ref (*8*).
d. Unrestricted Hartree-Fock, ref (*9*).
e. Restricted open-shell Hartree-Fock, ref (*10*).
f. Generalized valence bond, ref (*11*).
g. Multi-configuration SCF, ref (*12*).

B. Model, communication software, and general ideas

The single-program, multi-data (SPMD) model is used in the parallelization of GAMESS with each node executing essentially the same code. This model has many advantages for a large FORTRAN program (over 120,000 lines of code). One is that only one code needs to be maintained. Another advantage is that it is relatively easy to parallelize new sections of the code, since only one code needs to be examined for parallel content. In the early stages of the parallelization of GAMESS, only certain portions of the code were allowed to run in parallel. An error message would be given to a user who tried to run parallel jobs on sections of the code that were not parallelized and then the job would abort. As furthur portions of the code were parallelized, the error messages were removed.

An important consideration when parallelizing any code is which communication software package to use. Several criteria had to be met for GAMESS. First, portable software was needed since GAMESS executes on many different platforms. Second, software that required only a small learning curve was needed in order to facilitate the process, since the objective is to parallelize quantum chemistry codes, not necessarily to become experts in parallel communication. Third, the communication software had to work with quantum chemistry codes (i.e. usable with FORTRAN). Finally, the software must be either free or very cheap so that any user could obtain it. Several software packages were available at the beginning of our research, but the one that fit the above criteria best was the TCGMSG package of Harrison (*13*). This code is portable across several different platforms including UNIX workstations connected by Ethernet, distributed memory machines such as the Intel Paragon and shared memory machines such as the Alliant. Further, only about a dozen functions and subroutines are needed to perform the majority of the communications. Global functions are available to perform many of the operations, such as global summations of vectors and broadcasting a message from one node to all nodes. TCGMSG was specifically written to work with chemistry codes. And finally, TCGMSG is available via anonymous ftp, and therefore it is available to essentially all interested users.

Once the communication software and the model of parallelization are chosen, the "real" parallelization work can begin. First, the program should be relatively up to date before it is parallelized. It is not, in general, practical or useful to parallelize obsolete or very slow code. Also, direct methods tend to be easier to parallelize (at least at the first implementation level) than disk based methods since parallel disk I/O generally takes extra work to set up. Because of this, a direct method was introduced into the SCF code before the parallelization was initiated. Before development of the parallel MCSCF code, a faster transformation with direct capabilities (*14b*) was implemented.

One general consideration for any parallel code, is how I/O will be done. In GAMESS, only one node, the "master" node, reads input from the input deck and sends results to the output file. This requires that the master node "broadcast" input information to the other nodes. So, as an initial step in the parallelization of GAMESS, general I/O (as opposed to integral files, etc.) was made to execute only on the master for the entire code. This step actually consumed quite a bit of time, but in the end it proved to be very useful to have all of this work done at one time instead of working on it in small portions.

At this point, it is important to understand how the serial code actually works and what the computational bottlenecks are. Others(*1*) have identified the computational bottlenecks for SCF energies and gradients to be the computation of the two-electron integrals and two-electron gradient integrals, respectively. These investigators have developed methods for the parallelization of these parts of the code. In the end, of course, one wants as much of the code to run in parallel as possible (i.e., consider Amdahl's Law), but it is useful to attack the computational bottlenecks first. As part of the understanding of the serial code, it was useful to outline the actual subroutine calls made in GAMESS. By systematically examining the code, it was relatively easy to see which parts of the code could be parallelized. For example, even though it is not a computational bottleneck, the one- electron integrals can be parallelized in a manner similar to the two-electron integrals in very little programmer time. The actual parallelization of the SCF code is briefly outlined later in this paper.

During the parallelization process, it became apparent that at least two different methods of load balancing would be needed to obtain "good" efficiencies across many different platforms. The two methods used throughout GAMESS are called LOOP and NXTVAL load balancing. LOOP balancing is a static method that distributes the work by allowing each node to compute every mth block of work and skip the rest, resulting in an even distribution of many small pieces of work. This type of load balancing works best when the processors are of the same speed and have the same work load. The other type of load balancing, NXTVAL, is a dynamic algorithm using a shared counter which is managed by TCGMSG. This algorithm has each node send a message to the counter to get a new piece of work when it has finished its current work. The pieces of work must be of a relatively large size to overcome the cost of communicating with the shared counter. This algorithm works best when the processors are not of the same speed or do not have the sam work load.

During the parallelization process, several concepts were useful. One of these is the idea of global broadcast. For a global broadcast, one node has information that the rest of the nodes needs. This is the concept used when the "master" node sends input information to the other nodes. However, it can also be useful if one node performs a part of the calculation that the others do not and needs to broadcast the information to the other nodes. So, if one part of the calculation is found to operate more efficiently on one node than on several nodes (perhaps because the amount of communication would be greater than the amount of computation), one node can perform the calculation and broadcast the results to the other nodes. By using global communications in the code, the implementor does not need to worry about point to point communication, because the function supplied by the communication software (TCGMSG in this case) handles that for each type of hardware. Point to point communication may be needed in some cases, but in GAMESS, only global broadcasts are used.

Another important concept is the use of global summations. For example, in the current implementation of parallel SCF, each node calculates a partial contribution to the Fock matrix and then a global summation is performed. After the global summation, each node has the complete Fock matrix. An important point to remember is that each node must zero out the Fock matrix before it calculates its contribution, because the entire Fock matrix is summed. In other words, the global summation routine essentially gets the entire Fock matrix from each node, sums the pieces of the Fock matrix, and sends the result to each node. Related to the initial zeroing of matrices, occasionally vectors should be scaled before they are summed together. An example of this is in the gradient code. The one-electron gradient is calculated in parallel, globally summed, and written out to disk. The last step is performed for restart capabilities. When the two-electron gradient contribution is calculated, first the one-electron gradient is read from disk and the two-electron contributions are added. Since the one-electron gradient is completely self-contained, it must be divided by the number of processors so that the final result after the global summation of the two-electron gradient terms (which are calculated in parallel) is correct. Again, global summations are used wherever possible, instead of point to point communication, under the assumption that the global summations will be optimized by the communication software. (As will be seen in the MCSCF section, sparse vectors and matrices should NOT be globally summed to avoid wasting bandwidth.)

Another useful concept comes into play when debugging parallel code. While parallel debuggers are available, they are generally hard to use and can give misleading information. Debugging parallel code can be quite difficult, because the condition that results in an error is not always reproducable. However, when an error occurs frequently, a more systematic search for the error can be undertaken. We have found that flushing output for all the nodes and then aborting the job is a useful way to determine where the job is going wrong. This must be done at several places in the code. A place where everything is performing correctly needs to be identified (this is not necessarily as easy as it sounds). Then, a location where the error has already occured must be found. Then, it is a matter of printing out information from all of the nodes in between the two points and moving the abort as far down into the code as possible before detecting the error. The abort is very important, since it stops all activity of all of the nodes and can help to determine which nodes are failing where.

Another important tool used in the parallelization of GAMESS was stub routines. When the code is run sequentially, these stub routines are linked to the code instead of TCGMSG producing a serial version. However, there are some machines that TCGMSG has not yet been ported to or that have native functions identical to or comparable to the TCGMSG calls. Instead of porting TCGMSG to this machine, the appropriate calls were put into the stub routines, which then function as a translator betweeen TCGMSG and the native system calls. This isolates the machine specific code into only one source module that needs to be modified for a machine for which TCGMSG is unavailable or less efficient.

Finally, it should be noted that the approach described in the following paragraphs has advantages and disadvantages. It is likely that the SCF part of the code can be made to scale very well for large numbers of processors, as long as the size of the problem is scaled accordingly. At present the scalability of the analytic hessian and MCSCF codes is probably more limited, but even here there is a great benefit to users who have several workstations on which to run the parallel code. In addition, there are clear paths to improving the scalability of at least the analytic hessian code, and this is in progress. Since we have chosen to replicate the entire code on all nodes, each node must have sufficient memory to hold the larger executable.

C. SCF Parallelization

The specific details of the SCF code are given in reference *1a*. However, a general overview will be given here. The implementation of parallel SCF in GAMESS assumes that the Fock matrix and the density matrix are replicated on each of the nodes, instead of being distributed across the nodes. This limits the number of basis functions to around 400 on machines (such as the Intel Delta) with only 16 MB of memory per node and no virtual memory capabilities. This may seem to be a drastic limitation, but in practice, other issues become very important as the size of the problem increases. For example, for a modest basis set, such as 6-31G(d) (*15*), computations on relatively large molecules can be undertaken. One such example is the large cyclic adenosine monophsophate (cAMP) molecule with the molecular formula $C_{10}O_6N_5PH_{11}^-$. For a 6-31G(d) basis set, this molecule has 389 basis functions. For such a large molecule, finding the lowest energy conformation becomes a major challenge, not just because of the required computation time, but also due to the large number of conformations possible. So, even though a gradient may only take about 2 hours on 128 nodes of the Intel Delta, the intrinsic optimization problem will make finding the lowest energy conformer (or conformers) a daunting task. Nonetheless, it is important to explore the alternative of distributing the Fock matrix across the nodes as a means of increasing the size of the problems that can be tackled (*1*).

The following sections of the SCF code were modified to run in parallel: one-electron integrals, one-electron effective core potential (ECP) integrals, two-electron integrals, matrix multiplications, matrix diagonalization, one-electron gradient integrals, one-electron ECP gradient integrals and two-electron gradient integrals. The matrix diagonalization is actually only partially parallelized. When molecular symmetry is available in the molecule of interest, the Fock matrix is block diagonal. Each of these blocks can be sent to individual nodes to be diagonalized and then a global summation performed to get the total result on all nodes. When no molecular symmetry is available (i.e. C_1 symmetry), the diagonalization is completely serial and executes on only one node. The diagonalization step of an SCF calculation (order N^3, where N is the number of basis functions) actually becomes the bottleneck for a large enough problem, once the two-electron integrals (approximately of order N^4) have been parallelized (*11*). This means that the matrix diagonalization code needs to be a focus for new parallel developments. One approach for dealing with this bottleneck is to use a second-order method(*16*), but that has not yet been implemented into GAMESS. Details about the parallelization of the other steps in the SCF will not be given here since they have been given in many other studies and well accepted techniques were used.

Since the gradients are parallelized, optimizations, transition state searches, IRCs and numerical Hessians can also be executed in parallel. This provides the robustness of the parallel SCF part of the program. Many projects have already used the parallel SCF option of GAMESS to perform computations. Summaries of some of this work may be found in reference *17*.

D. Integral Transformation

One of the biggest challenges to the parallelization of post SCF and analytic Hessian codes is the AO to MO integral transformation(*3*). Formally, the transformation from AO ($<\mu\nu|\lambda\sigma>$) to MO ($<ij|kl>$) is an order N^5 operation

$$<ij||kl> = \Sigma_\mu C_{i\mu} \Sigma_\nu C_{j\nu} \Sigma_\lambda C_{k\lambda} \Sigma_\sigma C_{l\sigma} <\mu\nu|\lambda\sigma>$$

For all of the current applications in GAMESS, only a subset of the molecular integrals are needed. These are the $<ij|kl>$, $<aj|kl>$, $<ab|kl>$, $<aj|kb>$, and $<aj|bl>$ integrals, where i, j, k, l are MOs in the occupied space (core and active space for MCSCF, as discussed below), and a, b are MOs in the unoccupied (virtual) space. Since the transformation that was previously in GAMESS performed a full transformation, a new transformation (*14*) was incorporated into GAMESS. This transformation can use an unsorted list of AO integrals and molecular symmetry (Abelian groups only(*14b*)). The AO integrals can either be taken from disk or calculated directly. One of the options in this transformation performs passes over the full list of AO integrals to obtain subsets of the MO integrals. This algorithm is a perfect target for parallelization. Each node performs one or more passes over the AO integrals and obtains a subset of the MO integrals. In this way, the MO integrals are spread across all of the nodes and no communication is needed (unless NXTVAL load balancing is used). While this has the advantages of no communication and distributed MO integral storage, the algorithm also has the disadvantage that each node must either have the complete list of AO integrals available to it on disk or calculate the AO integrals each time they are needed. On a high communication speed parallel system, it should be possible to store only a subset of the AO integrals on each node, which can be broadcast to all nodes, so that each processes the entire AO list. We plan to implement this soon, since it will dramatically cut the AO integral storage requirement. Another potential disadvantage is that the number of passes must be evenly divisible by the number of processors (for LOOP load balancing); otherwise load imbalance can occur. However, the number of passes can be somewhat controlled by the amount of memory used for the transformation. So, in general, it is possible to ensure even load distribution. This transformation has now been interfaced to the MCSCF code, the CI code, the analytic Hessian code and the orbital localization code(*18d*). Parallelization of the CI code is explained below as part of the MCSCF implementation (*2a*).

New algorithms for the parallelization of transformations may well be designed in the future. However, new algorithms will still have the MO integrals distributed across the nodes. So, it would only be necessary to modify the interface code for the front end of a new transformation for it to work with the rest of GAMESS. At present, however, the algorithm described above works fairly well. This is especially true for the MCSCF calculation where the AO integrals (for a disk based method) are only calculated once for each MCSCF energy (which may involve approximately 10-20 iterations to obtain convergence). For more information about the transformation, the reader is referred to reference *2a*.

E. Approach to MCSCF

The parallelization of the MCSCF is presented in detail in reference *2a*, so only a brief overview will be given here. This reference also discusses the steps for the parallel CI code, an important part of the MCSCF code. First, some terms and issues must be discussed. Before an MCSCF wavefunction can be calculated (variously referred to as the full optimized reaction space (FORS) (*19*) or complete active space SCF (CASSCF) (*20*) formalism), the molecular orbitals must be partitioned into three different spaces. First, core orbitals with a fixed occupancy of two electrons must be identified. These orbitals generally do not contribute to the overall chemical reaction (i.e. they are not bond breaking or bond making orbitals). Next, an "active space" containing orbitals that are only partially occupied is identified. These are the orbitals that are directly involved in the chemical reaction and all possible configurations involving the active electrons and active orbitals are included in the calculation. Finally, the virtual or empty orbitals are identified.

A key step in an MCSCF calculation is the choice of starting orbitals. Usually, the active space in a FORS MCSCF calculation contains the orbitals corresponding to the bonds being broken and formed during some process of interest, the associated antibonding orbitals, and sometimes lone pairs that may play an important role in the process. Since this view of the active space is very chemical, a natural method for obtaining the starting orbitals is to make use of the localized orbital capabilities in GAMESS. The canonical molecular orbitals obtained directly from a Hartree-Fock calculation may be transformed (*18*) to more "chemical" localized molecular orbitals (LMO's) using well defined unitary transformations. In GAMESS this LMO transformation may be performed either on the complete set of valence orbitals or separately within each symmetry block. The advantage of the latter is that the preservation of symmetry minimizes the number of configuration state functions (CSF's) in the MCSCF calculation. The use of LMO's for choosing the active space makes it easy to identify the appropriate bonding MO's and lone pairs. In addition, it is a simple matter to reverse the phase of the bonding MO's to construct the corresponding antibonding orbitals needed to complete the active space. This is frequently a more effective procedure than using the canonical orbitals, since the canonical orbitals tend to be delocalized and therefore more difficult to identify as a particular antibonding moiety. Another effective choice for correlating orbitals are the modified virtual orbitals (*21*). These are derived from a cationic Fock operator, so they possess more valence antibonding character than the neutral virtual orbitals.

Another issue that must be discussed is the actual bottlenecks of the MCSCF calculation. Unlike the SCF code, the MCSCF has several different bottlenecks that depend on the type of calculation performed. For example, a molecule with only a few core orbitals and a relatively large active space will have the CI portion of the calculation as the bottleneck. On the other hand, a molecule with many core orbitals and a relatively small active space will have the transformation and the solution of the Newton-Raphson (NR) equations as bottlenecks. Therefore, it is imperative that as many of the steps as possible be parallelized. Because of limited space, only a brief discussion of the amount of parallelization in each step will be presented. For details, the reader should examine reference *2a*. The sections that are completely sequential are the initial orbital guess, calculation of the AO integrals, generation of the distinct row table, formation of the augmented orbital Hessian and the NR solutions. Of these, the first three are performed only once during the entire MCSCF energy calculation and formation of the augmented orbital Hessian is essentially trivial. However, solving the NR equations can be one of the major bottlenecks. The NR step is essentially a matrix diagonalization that finds the lowest eigenvector of the augmented Hessian. As mentioned earlier, parallel matrix diagonalizations are currently not very efficient (*1*).

Of the remaining steps in an MCSCF energy calculation, the molecular integral sort, the calculation of contributions (loops) to the CI Hamiltonian (*12*), the calculation of electron density matrices, formation of the Lagrangian and orbital Hessian (*22*) are only partially parallelized. The code for calculation of the contributions to the CI Hamiltonian and the electron density matrices have variable dependencies that are not easy to unravel, so essentially only disk I/O (distribution of loops across all disks) is parallelized. The other steps have global summations of large, relatively sparse matrices that require large amounts of communication. This communication time becomes comparable to or even larger than the CPU time savings from running in parallel. As mentioned earlier, these are probably places in the code where more care must be taken to send only the non-zero contributions, instead of the entire matrix. Finally, the integral transformation and the diagonalization of the Hamiltonian show very good efficiencies even with up to five RS6000/350 nodes tied together by Ethernet. The most time consuming part of the Davidson diagonalization is the formation of the Hamiltonian from the many loops distributed across all of the nodes. Each node forms a partial contribution to **HC**, after which a global sum is performed. Since I/O is performed in parallel, the scalability of this step is very good.

MCSCF gradients have also been parallelized so that actual chemical reactions can be explored using the parallel MCSCF technology. Specific timing examples and more detailed information for both the MCSCF energies and gradients can be obtained from references *2a, 17a*.

F. Analytic Hessians

Analytic Hessians in the MO basis involve several steps: (1) calculation of the AO integrals and the appropriate wavefunction; (2) transformation of integrals from the AO basis to the MO basis; (3) calculation of the one-electron second derivative (Hessian) integrals; (4) calculation of the two-electron Hessian integrals; and (5) solution of the coupled perturbed Hartree-Fock (CPHF) equations (*23*). Before a parallel transformation was available, a small scale algorithm was used in GAMESS (*4*). In that algorithm, all nodes would compute the one-electron Hessian integrals in parallel. Then, the master node performed the transformation while the other nodes (generally only 1-3 other nodes) calculated the two-electron Hessian integrals in parallel. After the master node finished with the transformation, it could participate in the calculation of the two-electron Hessian integrals if any were left to calculate. After steps 1-4 were finished, only the master node would complete the calculation by solving the CPHF equations.

Now that a parallel transformation is available, steps 1-4 can be performed in parallel. However, the full AO integral list must be calculated on each node and put onto a local disk (if using the disk based method) so that the parallel transformation works properly. This is an extra step that is not needed when the code is executed sequentially. Unfortunately, most of the CPHF solution is still performed sequentially and, after approximately 3 nodes, this becomes the computational bottleneck. Only the I/O to form the various pieces needed to set up the CPHF equations is performed in parallel. Since the matrices involved are quite large, the global summation takes essentially all of the time saved by the parallel I/O. This algorithm is the only one currently available in GAMESS.

As mentioned earlier, it is useful to make sure that the sequential code is relatively up to date before parallelization. The current method for solving the CPHF equations is relatively slow, so before an effort is made to parallelize this step, a new solver will be implemented. Using the same example used when the first analytic Hessian algorithm was published, Table II compares the computational times (on the master node) for each of the two algorithms. The test case is the C_s molecule 5-aza-2,8-dioxa-1-stibabicyclo[3.3.0]octa-2,4,6-triene ($SbO_2NC_4H_4$) using a 3-21G* basis set (*24*) giving 110 basis functions. The calculations were performed on three RS6000/350s dedicated to the test. The Ethernet connecting the three machines was not dedicated to the test and therefore, the tests had to compete with other packets on the network.

As can be seen in Table II, the new transformation is faster than the old one for one node. Also, the CPHF solution for the new algorithm is faster on one node than is the old algorithm. The actual CPHF code has not been changed. The difference in time comes from the number of integrals that must be read in and used. In the old algorithm, many of the integrals that were read in were discarded since the old algorithm performed a full transformation. The new transformation calculates only those integrals that are actually needed. The timing example also shows that indeed the CPHF step is the main bottleneck at three nodes. It is very clear that parallelization of the CPHF solution is needed before the scalability of the analytic Hessians can proceed to more nodes, but it is likely that good speedups on up to dozens of nodes will be achieved eventually.

Table II. Timing example. Time in seconds for the master using the new/old algorithm.

p=	1	2	3
	---	---	---
setup	0.57/0.58	0.69/0.78	0.73/0.84
1e- ints	1.10/1.12	0.87/0.86	0.88/0.84
huckel guess	15.77/15.77	15.74/16.46	16.17/16.96
2e- ints[a]	111.19/133.90	55.34/62.41	37.42/39.48
SCF cycles[b]	223.13/190.87	103.26/103.92	79.44/66.25
properties	2.23/1.61	2.46/2.44	2.63/2.78
2e- ints	-- /206.23	111.28/211.29	110.97/213.38
transformation[c]	1113.67/1881.05[d]	552.38/1902.15	381.09/1897.92
1e- hess ints	28.20/28.62	16.46/17.05	14.63/14.74
2e- hess ints	3322.92/3367.57	1668.86/83.93	1113.37/12.41
CPHF	1438.66/1653.75	1433.34/1673.50	1477.32/1664.48
	-------	------	------
total CPU	6258.01/7481.69	3961.34/4075.05	3235.27/3930.85

a. 6,125,653 AO integrals.
b. 13 iterations.
c. 5,871,750 MO integrals.
d. This time includes the integral ordering needed in the old transformation as well as the actual transformation time.

G. Parallel MP2 Code

A new MP2 code from HONDO (*14*) has been incorporated into GAMESS. Since this code had already been parallelized, the parallel calls in the new code were translated to TCGMSG. A brief description of the HONDO algorithm used will be presented here.

The MP2 code includes its own specialized transformation. If the AO integrals are calculated directly, the MP2 transformation is essentially the same as the one described earlier in this paper. However, the disk based transformation works differently. The MP2 transformation assumes that the AO integrals are distributed across all of the nodes. Each node is assigned a range of MO integrals to calculate. Then, each node (in its turn) reads in a buffer load of integrals, broadcasts the buffer to all other nodes, and calculates the contributions of those AO integrals to its range of MO integrals. When all of the AO integrals from every node have been used, the MO integrals are used to form contributions to the MP2 energy. Thus, the actual MO integrals are not sent to disk, only held in memory. If the nodes cannot hold their ranges of MO integrals in memory, several passes over the AO integrals are needed. This transformation has the advantage that it uses the AOs as they are distributed across the nodes (i.e. the entire AO list does not need to be calculated on each node). However, it does require the broadcast of order N^4 AO integrals resulting in a large amount of communication. This particular implementation of the algorithm also has the disadvantage that it is only implemented for RHF wavefunctions.

H. Conclusions

Most of the functionality of GAMESS now executes in parallel. Table III provides a summary of the parallel portions of GAMESS.

While most of GAMESS has been parallelized, further optimizations of existing algorithms and better algorithms are needed to improve the general efficiencies of the code. Specifically, the following areas need more work: parallel matrix diagonalizations (directly affecting the SCF and NR solution in the MCSCF); large global summations of large sparse matrices need to be made much more efficient; solution of the CPHF equations in parallel; and new parallel transformations are needed.

Table III: Tabulated overview of parallel GAMESS

	Energy	Gradient	Hessian[a]	MP2 CI	Semi
RHF	x	x	x	x	x
UHF	x	x			
ROHF	x	x	x		x
GVB	x	x	x		x
MCSCF	x	x			x

a. Refers to analytic evaluation. Numerical Hessians are available whenever analytic gradients are available.

Acknowledgement

The authors are grateful to Dr. Michel Dupuis for permission to incorporate his serial partial integral transformation code and his parallel RHF MP2 program into GAMESS. This work was supported by a grant from the Air Force Office of Scientific Research (93-1-0105), a Department of Education GAANN fellowship to TLW and with the assistance of the Advanced Research Projects Agency. The IBM RS6000s were provided by Iowa State University.

References

1.(a) Schmidt, M.W.; Baldridge, K.K.; Boatz, J.A.; Elbert, S.T.; Gordon, M.S.; Jensen, J.H., Koseki, S.; Matsunaga, N., Nguyen, K.A.; Su, S.; Windus, T.L.; Dupuis, M.; Montgomery Jr., J.A. *J.Comput.Chem.* **1993**, *14*, 1347; (b) Dupuis, M.; Watts, J.D.

Theor.Chim.Acta **1987**, *71*, 91; (c) Guest, M.F.; Harrison, R.J.; van Lenthe, J.H.; van Corler, L.C.H. *Theor.Chim.Acta* **1987**, *71*, 117; (d) Ernenwein, R.; Rohmer, M.M.; Benard, M. *Comput.Phys.Comm.* **1990**, *58*,305; (e) Rohmer, M.M.; Demuynck, J.; Benard, M.; Wiest, R.; Bachmann, C.; Henriet, C.; Ernenwein, R. *Comput.Phys.Comm.* **1990**, *60*, 127; (f) Harrison, R.J.; Kendall, R.A. *Theor.Chim.Acta* **1991**, *79*, 337; (g) Cooper, M.D.; Hillier, I.H. *J.Computer-Aided Molecular Design* **1991**, *5*, 171; (h) Kindermann, S.; Michel, E.; Otto, P. *J.Comput.Chem.* **1992**, *13*, 414; (i) Luthi, H.P.; Mertz, J.E.; Feyereisen, M.W.; Almlof, J.E. *J.Comput.Chem.* **1992**, *13*, 160; (j) Brode, S.; Horn, H.; Ehrig, M.; Moldrup, D.; Rice, J.E.; Ahlrichs, R. *J.Comput.Chem.* **1993**, *14*, 1142; (k) Feyereisen, M.; Kendall, R.A. *Theor.Chim.Acta* **1993**, *84*, 289; (l) Colvin, M.E.; Janssen, C.L.; Whiteside, R..A.; Tong, C.H. *Theor. Chim. Acta* **1993**, *84*, 301; (m) Also see *Theor.Chim.Acta* **1993**, *84* (No. 4-5), 255ff.

2. (a) Windus, T.L.; Schmidt, M.W.; Gordon, M.S. *Theor.Chim.Acta* in press; (b) Watts, J.D.; Dupuis, M. *J.Comput.Chem.* **1988**, *9*, 158; (c) Rendall, A.P.; Lee, T.J.; Lindh, R. *Chem.Phys.Lett.* **1992**, *194*, 84; (d) Harrison, R.J. *J.Chem.Phys.* **1991**, *94*, 5021; (e) Harrison, R.J.; Stahlberg, E. *CSCC Update* **1992**, *13*, 5; (f) Schuller, M.; Kovar, T.; Lischka, H.; Shepard, R.; Harrison, R.J. *Theor.Chim.Acta* **1993**, *84*, 489.

3.(a) Whiteside, R.A.; Binkley, J.S.; Colvin, M.E.; Schaefer III, H.F. *J.Chem.Phys.* **1987**, *86*, 2185; (b) Covick, L.A.; Sando, K.M. *J.Comput.Chem.* **1990**, *11*, 1151; (c) Wiest, R.; Demuynck, J.; Benard, M.; Rohmer, M.M.; Ernenwein,R. *Comput.Phys.Commun.* **1991**, *62*, 107; (d) Limaye, A.C.; Gadre, S.R. *J.Chem.Phys.* **1994**, *100*, 1303.

4. Windus, T.L.; Schmidt, M.W.; Gordon, M.S. *Chem.Phys.Lett.* **1993**, *216*, 375.

5.(a) Pople, J.A.; Binkley, J.S.; Seeger, R. *Int.J.Quantum Chem.* **1976**, *10*, 1; (b) Carsky, P.; Hess, B.A.; Schaad, L.J. *J.Comput.Chem.* **1984**, *5*, 280.

6 (a) Brooks, B.; Schaefer III, H.F. *J.Chem.Phys.* **1979**, *70*, 5092; (b) Brooks, B.; Laidig, W.; Saxe, P.; Handy, N.; Schaefer III, H.F. *Physica Scripta* **1980**, *21*, 312.

7. Stewart, J.J.P. *J.Computer-Aided Mol.Design* **1990**, *4*, 1.

8. Roothaan, C.C.J. *Rev.Mod.Phys.* **1951**, *23*, 69.

9. Pople, J.A.; Nesbet, R.K. *J.Chem.Phys.* **1954**, *22*, 571.

10. Guest, M.F.; Saunders, V.R. *Mol.Phys.* **1974**, *28*, 819.

11. Bobrowicz, F.W.; Goddard, W.A. *Modern Theoretical Chemistry*, H.F. Schaefer, Ed., Plenum Press, New York, **1977**, p. 79.

12. General review: Roos, B.O. Methods in *Computational Molecular Physics*. G.H.F. Diercksen, S. Windson, Eds., D. Reidel Publishing, Dordrecht, Holland, **1983**, p.161.

13.(a) Harrison, R.J. *Int.J.Quantum Chem.*, **1991**, *40*, 847; (b) Harrison, R.J. version 4.0.2. is available by anonymous ftp in directory /pub/tcgmsg from host ftp.tcg.anl.gov.

14.(a) Dupuis, M.; Chin, S.; Marquez, M. *Relativistic and Electron Correlation Effects in Molecules and Clusters*, G.L. Malli, Ed., NATO ASI Series, Plenum Press, New York, **1992**; (b) Hollauer, E.; Dupuis, M. J.Chem.Phys. **1992**, *96*, 5220.

15.(a) Ditchfield, R.; Hehre, W.J.; Pople, J.A. *J.Chem.Phys.* **1971**, *54*, 724; (b) Hehre, W.J.; Ditchfield, R.; Pople, J.A. *J.Chem.Phys.* **1972**, *56*, 2257.

16. Shepard, R. *Theor. Chim. Acta* **1993**, *84*, 343.

17.(a) Windus, T.L.; Schmidt, M.W.; Gordon, M.S. *Mardi Gras Symposium* **1994**, submitted; (b) Schmidt, M.W.; Windus, T.L.; Jensen, J.; Matsunaga, N.; Cundari, T.R.; Boatz, J.A.; Baldridge, K.K.; Gordon, M.S. *ACS Symp. Series on Parallel Computing* **1994**, in preparation.

18.(a) Boys, S.F. *Quantum Theory of Atoms, Molecules, and Solids*, P.O. Lowdin, Ed., Academic Press, New York, **1966,** p. 253; (b) Kleier, D.A.; Halgren, T.A.; Hall, J.H.; Lipscomb, W.N. *J.Chem.Phys* **1974,** *61,* 3905; (c) Pipek, J.; Mezey, P.G. *J.Chem.Phys.* **1989,** *90,* 4916; (d) Edmiston, C.; Ruedenberg, K. *Rev.Mod.Phys.* **1963,** *35,* 457.

19.(a) Ruedenberg, K.; Schmidt, M.W.; Gilbert, M.M.; Elbert, S.T. *Chem.Phys.* **1982,** *71,* 41; (b) Ruedenberg, K.; Schmidt, M.W.; Gilbert, M.M. *Chem.Phys.* **1982,** *71,* 51; (c) Ruedenberg, K.; Schmidt, M.W.; Gilbert, M.M.; Elbert, S.T. *Chem.Phys.* **1982,** *71,* 65.

20. Siegbahn, P.E.M.; Almlof, J.; Heiberg, A.; Roos, B.O. *J.Chem.Phys.* **1981,** *74,* 2384.

21. Bauschlicher, C.W. *J. Chem. Phys.* **1980,** *72,* 880.

22. Siegbahn, P.; Heiberg, A.; Roos, B.; Levy, B. *Physica Scripta* **1980,** *21,* 323.

23.(a) King, H.F.; Komornicki, A. *Geometrical Derivatives of Energy Surfaces*, P. Jorgenson, J. Simon, Eds. NATO ASI Ser. C., *Vol. 166,* Reidel Publishing, Dordrecht, **1986,** p. 207; (b) Osamura, Y.; Yamaguchi, Y.; Saxe, P.; Fox, D.J.; Vincent, M.A.; Schaefer III, H.F. *J.Mol.Struct.* **1983,** *103,* 183; (c) Duran, M.; Yamaguchi, Y.; Schaefer III, H.F. *J.Phys.Chem.* **1988,** *92,* 3070.

24.(a) Binkley, J.S.; Pople, J.A.; Hehre, W.J. *J.Am.Chem.Soc.* **1980,** *102,* 939; (b) Dobbs, K.D.; Hehre, W.J. *J.Comput.Chem.* **1986,** *7,* 359.

RECEIVED November 15, 1994

Chapter 3

Applications of Parallel GAMESS

Kim K. Baldridge[1], Mark S. Gordon[2], Jan H. Jensen[2],
Nikita Matsunaga[2], Michael W. Schmidt[2], Theresa L. Windus[3],
Jerry A. Boatz[4], and Thomas R. Cundari[5]

[1]San Diego Supercomputer Center, P.O. Box 85608, San Diego, CA 92186
[2]Department of Chemistry, Iowa State University, Ames, IA 50011
[3]Department of Chemistry, Northwestern University, Evanston, IL 60208
[4]Phillips Laboratory, OLAC PL/RKFE,
Edwards Air Force Base, CA 93523
[5]Department of Chemistry, University of Memphis, Memphis, TN 38152

In this paper we discuss several recent applications that would have been difficult or impossible without the availability of the parallel implementation of the electronic structure code GAMESS. These applications include the study of highly strained rings, such as inorganic prismanes and bicyclobutanes, cage compounds such as cyclophanes and atranes, the neutral <-> zwitterion isomerization of glycine, transition metal-main group binding, and the implementation of parallel graphics.

The previous paper presented an outline of the strategy used in converting the electronic structure code GAMESS to a general parallel code [1]. In this paper, we turn to a brief discussion of several applications of this code. The parallel capability of GAMESS has already been used to solve a broad spectrum of problems of importance to organic, inorganic, organometallic, and biochemistry that would otherwise have been impossible within a reasonable time frame. Indeed, the parallel capability allows us to perform calculations on relevant compounds within a time frame that is meaningful to experimental colleagues. Several of these applications are summarized in the following sections.

I. Highly Strained Rings

An important area of application for parallel GAMESS has been the design of metastable species that have potential as new high energy fuels or fuel additives. Two such endeavors have been the study of the potential energy surface of the BN analog of prismane and the tetrasila analog of bicyclobutane.

A. BN Prismanes
An example of the performance (and the difficulties) of parallel SCF calculations is provided by the BN analog of prismane,

0097–6156/95/0592–0029$12.00/0
© 1995 American Chemical Society

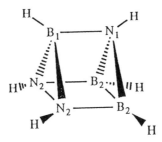

a high energy isomer of the benzene analog borazine. This relatively small example (169 basis functions) serves to illustrate some of the successes and potential bottlenecks that arise from parallel computations. The results are summarized in **Table 1**, where the speedups for an energy plus gradient run are presented as a function of the number of processors. The overall speedup (last column) is essentially perfect (100%) through 8 processors, 92% through 16 processors, and slowly tails off as the number of processors increases to 256. Even at 256 processors there is a better than 25% speedup, and one expects that as the size of the problem is increased, the tailing off of efficiency will occur more slowly. The source of the loss in efficiency as the number of processors increases may be determined by analyzing the middle three columns of the table. While the two-electron gradients are essentially perfectly parallel, the efficiency of the Hartree-Fock part of the calculation parallels that of the overall job. Further analysis reveals that, while the calculation is dominated by the (almost perfectly parallel) integrals plus gradients for small numbers of processors, the sequential Fock matrix diagonalization becomes a larger percentage of the calculation as the number of processors is increased. Since matrix diagonalizations are such an important part of electronic structure calculations, an effective treatment of this part of the calculation in parallel computations must be addressed.

Table 1. Incremental Performance Advantage and Efficiency of BN Prismane[a]

# Proc.	int + RHF	1e⁻ grad	2e⁻ grad	total
8	1.0	1.0	1.0	1.0
16	1.87(93.5)	1.89(94.5)	1.98(99.0)	1.84(92.0)
32	3.26(81.5)	1.56(39.0)	3.91(97.8)	3.09(77.2)
64	5.22(65.2)	1.59(19.9)	7.64(95.5)	4.74(59.2)
128	7.77(48.6)	3.19(19.9)	4.74(92.1)	6.74(42.1)
256	10.02(31.3)	3.36(10.5)	27.32(85.4)	8.35(26.1)

[a]The values in parentheses are efficiencies. The calculations are carried out at the RHF/6-311G(2d,2p)//RHF/SBK(d) level of theory (204 basis functions) on a 512-node Intel's Touchstone Delta (16Mb memory/node).

The known potential energy surface for BN prismane is shown in **Figure 1**. At the MP2/SBK(d)//RHF/SBK(d) level of theory, BN prismane is 161 kcal/mol higher in energy than the borazine global minimum [2]. Several other minima have been found on this surface, including a planar isomer of borazine that is itself 100 kcal/mol higher in energy than borazene. To date, no direct route from BN prismane to borazine has been found, and all routes leading from BN prismane appear to involve energy barriers in the range of 30-40 kcal/mol. Of particular interest is the pair of three-membered rings shown at the right in the figure. Since these rings lie 40 kcal/mol *above* BN prismane, they may provide a synthetic route to this high energy species. Potential syntheses are being explored at Rockwell Science.

B. Tetrasilabicyclobutanes

The tetrasila-analog of bicyclobutane has been of interest for several years, since it is predicted [3] by electronic structure theory to exist as two isomers (**Figure 2**) that differ primarily in the length of the bridgehead Si-Si distance, a normal 2.35Å in the short bond (SB) isomer and a much longer 2.9Å in the long bond (LB) isomer. In the unsubstituted compound, the LB isomer is predicted to be lower in energy by about 10 kcal/mol, at the GVB/6-31G(d) level of theory. The only analog that has been synthesized is highly substituted, with t-butyl groups replacing the hydrogens at both bridgehead positions and substituted phenyl rings at the peripheral positions. In contrast to the theoretical predictions, only the SB isomer is found for this substituted compound. This difference between theory and experiment is important to understand. The unsubstituted compound (which has not yet been synthesized), may be used as an additive to the most common propellant used in space launches: liquid oxygen (LOX)/liquid hydrogen (LH2) mixtures. Using 2.5 mole % of the unsubstituted compound in LOX/LH2 is found to increase the specific impulse (I_{sp}, the most common measure of fuel effectiveness, is proportional to the energy gain and inversely proportional to the mass of the combustion products) by 11 seconds. This translates into a savings of several million dollars per launch. Therefore, GVB/6-31G(d) calculations were performed on the SB->LB isomerization as a function of the group R in the bridgehead positions. As shown in **Table 2**, increasing the size of the bridgehead substituents, destabilizes the LB isomer, relative to the SB isomer. This explains why only the SB isomer is found experimentally. Of particular interest is our prediction that the two isomers of the dimethyl analog are nearly isoenergetic, separated by about a 6 kcal/mol barrier. This suggests that both isomers of the dimethyl compound may be synthesized. Note that the size of the basis set for the di-t-butyl species (nearly 200 basis functions) necessitated the use of parallel GAMESS for the timely completion of the project.

II. Cage Compounds

We have recently been interested in a series of cage compounds that may be generally represented as

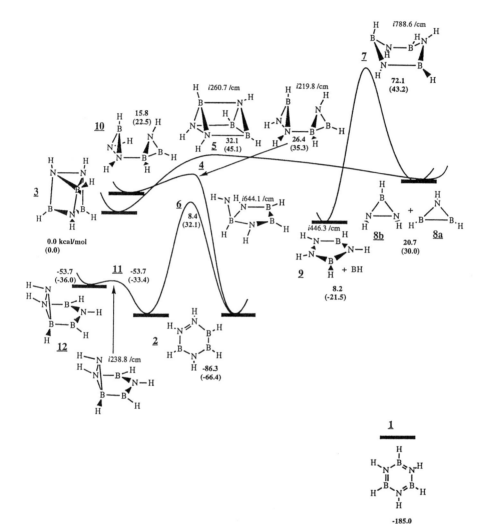

Figure 1. RHF/SBK(d) Potential Energy Surface of BN Prismane.
The values are in kcal/mol. The values in parentheses are of MP2
relative energies. All ZPE's are corrected by multiplying by 0.89.

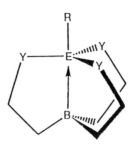

With the apex E atom = Si, P, Ti, etc., they are silatranes, phosphatranes, titanatranes, etc. The impact of the equatorial atoms Y attached to E (e.g., Y = N are aza-atranes) and the axial substituents R on the chemical and physical properties of the atranes is of considerable interest. Silatranes, for example, are precursors for new SiO materials, while the phosphatranes are strongly basic compunds with great potential as catalysts. When the base of the cage compounds is a benzene ring, with the three methylene strands attached at the 1,3,5 positions, these compounds are cyclophanes.

A. Cyclophanes

Pascal has synthesized the carbon cyclophane, with the bridgehead H pointing inside (ENDO) the cyclophane ring [4]. This unusual geometric arrangement prompted (unsuccessful) experimental efforts to synthesize the sila-analog. In an attempt to understand why (apparently) the ENDO structure is preferred in the carbon compound, whereas the EXO structure is preferred in the silicon compound, we performed a series of RHF/6-31G(d) calculations on both C and Si cyclophanes [5]. The *ab initio* calculations predict that the ENDO isomer is 13 kcal/mol lower in energy than EXO, in agreement with Pascal's experiments. Replacing the apex carbon with a silicon results in a dramatic reversal of stability, with the EXO structure now preferred by 43 kcal/mol! Again, this is consistent with the synthetic difficulties for this species. A simple explanation for this lies in the bond dipoles of C-H vs. Si-H. Whereas the former bond is polarized C^- H^+, the latter is polarized Si^+H^-. So, in the carbon case one has a positively charged H pointing towards the negative benzene π cloud, whereas in the silicon compound it is a negative H that points toward the π cloud. So, for C it is an attractive interaction, while for Si the interaction is repulsive. This assertion may be assessed in two ways. One might consider replacing the H at the bridgehead by a more electropositive element, such as Li. Doing so makes the ENDO structure only slightly (3 kcal/mol) more favorable for the C case, but stabilizes the ENDO structure by 40 kcal/mol for Si! One might also enlarge the cage from two to three carbons/strand, to alleviate the repulsive interaction and the internal crowding. For the carbon compound, the larger cage favors the ENDO structure by 18 kcal/mol, as compared with 13 kcal/mol for the smaller cage. For the silicon species, the larger cage favors EXO by only 2.5 kcal/mol, compared with 43 kcal/mol for the smaller cage. These extensive calculations, made possible by parallel electronic structure codes, therefore predict that it may be possible to synthesize the ENDO structure by making the cage one carbon larger.

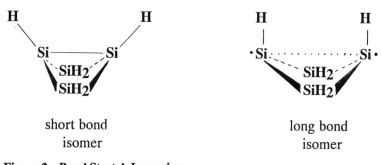

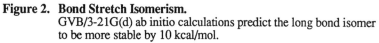

short bond   long bond
isomer   isomer

Figure 2. Bond Stretch Isomerism.
GVB/3-21G(d) ab initio calculations predict the long bond isomer
to be more stable by 10 kcal/mol.

Table 2.   Relative Energies of $R_2Si_4H_4$ Bond-Stretch Isomers [a]

	R = H		
Level of theory	SB	TS	LB
GVB/3-21G*//GVB/3-21G*	0.0 (0.0)	2.1 (2.0)	-9.6 (-9.7)
GVB/6-31G(d)//GVB/3-21G*	0.0 (0.0)	1.1 (1.0)	-12.4 (-12.5)
SOCI/6-31G(d)//GVB/3-21G*	0.0 (0.0)	1.3 (1.1)	-10.1 (-10.1)

	R = CH_3		
	SB	TS	LB
GVB/3-21G*//GVB/3-21G*	0.0 (0.0)	6.3 (6.3)	-1.9 (-2.0)
GVB/6-31G(d)//GVB/3-21G*	0.0 (0.0)	4.5 (4.5)	-4.5 (-4.6)
SOCI/6-31G(d)//GVB/3-21G*	0.0 (0.0)	5.1 (5.1)	-1.8 (-1.9)

	R = $C(CH_3)_3$		
	SB	TS	LB
GVB/3-21G*//GVB/3-21G*	0.0 (0.0)	7.2 (6.9)	4.9 (4.6)
GVB/6-31G(d)//GVB/3-21G*	0.0 (0.0)	6.0 (5.7)	2.5 (2.2)
SOCI/6-31G(d)//GVB/3-21G*	0.0 (0.0)	6.8 (6.5)	5.1 (4.8)

[a] Energies in kcal/mol. ZPE-corrected energies in parentheses.

B. Silatranes

In the solid state, most silatrane trans-annular (SiN_t) distances are 2.05 - 2.20Å. This is considerably shorter than the sum of the van der Waals radii (3.5Å), but much longer than typical SiN single bond lengths (1.7-1.8Å). These relatively weak SiN_t bonds are even longer in the gas phase. The two gas phase structures (for R = F[6] and CH_3[7]) reveal SiN_t distances that are 0.28Å longer than those in the corresponding crystals. The solution phase SiN_t distances appear to be intermediate between those in the gas phase and solid state [8]. This indicates that the SiN_t bond is weak and easily deformed.

The silatrane series (E = Si) has been studied as a function of Y (= O, NH, NMe, CH_2) and R (= H, F, OH, NH_2, CH_3, Cl, SH, PH_2, SiH_3) [9]. Key issues are the fundamental nature of the transannular SiN_t interaction and the difference between gas and condensed phases. The geometry optimizations were performed at the SCF level of theory, mostly using the 6-31G(d) basis set. The general approach has been to obtain starting structures with semi-empirical AM1[10] or PM3[11] geometry optimizations and hessians. The *ab initio* calculation (run in direct mode) typically required 400-800 minutes of time on 128 nodes of the Intel Touchstone Delta at CalTech.

Because the SiN_t bond is so weak, it is difficult to accurately reproduce the experimental distances. At the RHF/6-31G(d) level of theory, the two known gas phase SiN_t distances are over-estimated by more than 0.2Å. Expanding the basis set to include two sets of d functions on Si and its five adjacent heavy atoms, plus a set of diffuse sp functions on the same six heavy atoms, decreases the SiN_t distance in the R = F silatrane by 0.12Å to 2.416Å, bringing it into much closer agreement with the experimental value of 2.32Å. The remaining error is due to correlation and additional basis set effects.

The softness of the SiN_t bond is dramatically illustrated by plotting the energy of the R = F silatrane as a function of the SiN_t distance. When this distance is varied over a 0.5Å range, the energy increases by only 4 kcal/mol! This means that crystal packing forces need not be larger than 1 kcal/mol to produce the observed 0.28Å compression of the silatrane SiN_t bond. The effect of condensed phase on the SiN_t distance has been investigated by modeling the effect of solvent DMSO with a simple reaction field cavity model [12]. A cavity radius of 3.67Å, derived from the experimental density of fluorosilatrane, and a dielectric constant of 45 were used in these simulations. Re-optimization of the geometry in the presence of solvent decreases the SiN_t distance by 0.31Å, essentially the difference between the experimental gas phase and crystal bond lengths! In addition to the softness of the SiN_t surface already discussed, this large change in the SiN_t distance is also due to the large dipole moment in the silatranes which interact strongly with the solvent dielectric.

Finally, Boys localized molecular orbitals (LMO's) [13] have been used to develop an understanding of the nature of the SiN_t bond. These LMO's illustrate that the SiN_t bonds are best described as nitrogen lone pairs, interacting only weakly with the transannular silicon atoms. This supports the interpretation of these bonds as dative in character, in agreement with the description put forth by Haaland [14].

C. Phosphatranes

The proton affinities of a series of azaphosphatranes (E = P, Y = N) have been studied to determine their relative base strengths and the effect of substitution and protonation on the transannular PN_t interaction [15]. As was the case for the silatranes discussed above the *ab initio* investigation of this large group of complex molecules (substituents Z on P = H^+, F^+, Cl^+, CH_2, CH_3^+, NH, NH_2^+, O, O^+) would have been impractical without the availability of parallel electronic structure codes and parallel computers. These calculations were performed on the Intel Touchstone Delta, in a manner analogous to that described above for the silatranes, using similar basis sets at the SCF level of theory. Four molecules, with Z = CH_2 or NH, have been found to be stronger bases than the parent compound, suggesting that these species are likely targets for new catalysts. Analysis of the PN_t distances and electron density analyses show that there is clear evidence for transannular dative bonding in the cationic species. Protonation clearly results in a dramatic strengthening of this bond. The corresponding bond distances decrease by more than 1Å! The use of the reaction field model to simulate the solvent DMSO suggests that the basicity trends found for the gas phase compounds are not changed in solution.

III. Glycine Isomerization

It is well known that in solution amino acids exist primarily as zwitterions (Z), whereas in the gas phase only the neutral structure (N) is a minimum on the potential energy surface (PES). It is not clear, however, what forces fundamentally operate to stabilize Z relative to N. One can, for example, simulate the bulk effects of aqueous solution using a simple reaction field model [16], and such calculations do predict the Z form to be more stable. However, such calculations do not address the role played by individual

electronic interactions between water molecules from the solvent and the amino acid. To explore this question, we have employed *ab initio* quantum chemistry to explore the effects on the N <-> Z equilibrium of successively adding water molecules to the simplest amino acid glycine. The geometry optimizations and subsequent tracing of the minimum energy paths (MEP's) were performed at the SCF level of theory, on the 16 node iPSC 860 Intel parallel computer, located at Kirtland AFB, using both the 6-31G(d) [17] and Dunning DZP [18] basis sets. Additional single point calculations were performed at both the MP2 and MP4 levels of theory [19], using the aforementioned basis sets, as well as the Dunning correlation consistent basis sets [20].

Krogh-Jespersen has recently demonstrated that Z is not a minimum on the isolated glycine PES [21], when adequate basis sets are used. The energetics for the isomerization of the glycine·H_2O complex are summarized in **Figure 3**. At the SCF level of theory minima are found at both the N and Z structures; however, the transition state separating the two isomers, and therefore the Z minimum, disappears at correlated levels of theory. Note that the water molecule does not directly particpate in the proton transfer. Rather, it functions as an observor, so we refer to this process as intramolecular proton transfer. It is important to note that the use of larger basis sets and higher levels of perturbation theory (i.e., MP4) have less than a 1 kcal/mol effect on the predicted energetics. The MP2/DZP++//SCF/DZP energy of the neutral glycine-water complex is 12.8 kcal/mol below that of the zwitterion when vibrational zero point (ZPE) corrections are included. The same Z structure shown in **Figure 3** can transfer a proton through the water molecule (water-assisted proton transfer), via a different transition state. The energetics for this process are shown in **Figure 4**. Unlike the intramolecular proton transfer, the transition state for the water-assisted proton transfer still exists upon the addition of larger basis sets and correlation corrections. Addition of ZPE corrections does raise Z above the transition state, and the same Z structure is unstable to isomerization via the intramolecular route. However, these results suggest that the water-assisted proton transfer may be the more viable way to stabilize Z in cases for which more than one Z·nH_2O isomer may exist.

The energetics for the intramolecular and water-assisted proton transfer mechanisms for the glycine·$2H_2O$ complex are shown in **Figures 5** and **6**, respectively. The features of the potential energy curve for the intramolecular route are similar to those for the single water complex. This Z complex is found to be a minimum at the SCF level of theory, but the transition state again disappears at correlated levels of theory. Unlike the single water case, the water-assisted route for glycine·$2H_2O$ originates from a different (essentially isoenergetic) Z isomer. This is important, because we find that the transition state for this route remains higher than both Z and N, even after the addition of correlation and ZPE corrections. It is also important that the Z isomer is found to be only 4.8 kcal/mol higher in energy than N, in the presence of two water molecues. It is reasonable to consider that part of the PES that connect the two glycine·$2H_2O$ zwitterion structures, corresponding to the intramolecular and water-assisted routes. The transition state that connects these two structures has been identified, and the barrier separating them is 8 kcal/mol at the MP2/DZP++ level of theory. So, the glycine·$2H_2O$ zwitterion appears to be a stable minimum on the PES. These results clearly demonstrate that electronic interactions between the amino acid and individual solvent molecules play a crucial role in the stabilization of the zwitterion species.

IV. Transition Metal Complexes

The transition metals (TMs) constitute a family of elements of importance in advanced materials, biochemistry, and catalysis.[22] The large size of many TM complexes and the demands of the methods needed to accurately describe their chemistry make parallel computing very attractive in this area. Our main algorithmic approach to the challenges of computational TM chemistry entails the design, testing and use of effective core

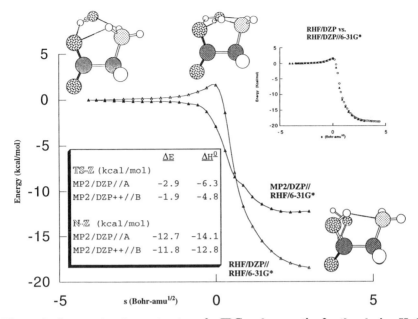

Figure 3. Intramolecular proton transfer IRC and energetics for the glycine-H₂O complex.

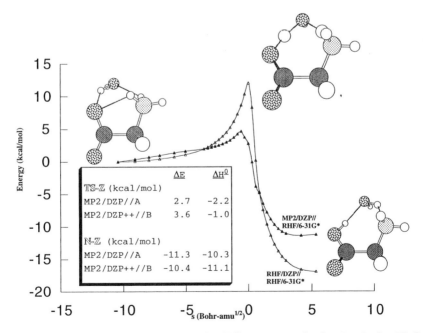

Figure 4. Water-assisted proton transfer IRC and energetics for the glycine-H₂O complex.

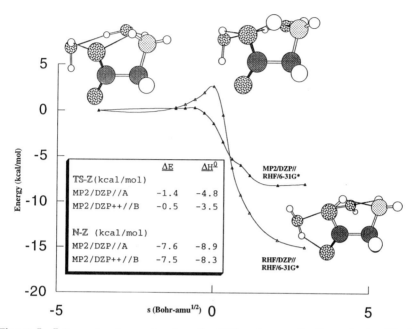

Figure 5. Intramolecular proton transfer IRC and energetics for a glycine·2H₂O complex.

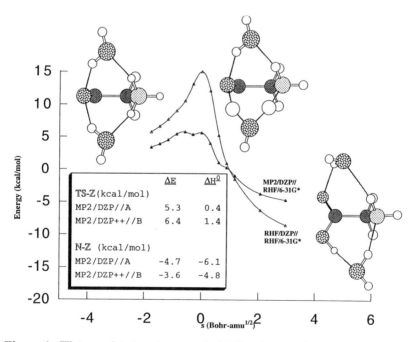

Figure 6. Water-assisted proton transfer IRC and energetics of a glycine·H2O complex.

potentials (ECPs).[23] It is important to compare the efficiency of parallel ECP codes with traditional all-electron methods. The challenges which arise in applying quantum chemical methods to TMs have been discussed in more detail previously.[23, 24] A representative problem is chosen below to illustrate the possibilities for parallel code in TM chemistry.

Recently there has been a great focus on complexes with multiple bonds between TMs and heavier main group (MG) elements.[25] Apart from a fundamental interest in multiple bonding involving heavier MG elements such complexes have been envisioned as precursors and intermediates in the synthesis of solid-state advanced materials.[25] An exciting series of TM=MG(heavy) complexes is provided by Parkin and Howard[25a,b], $Cp'_2M(E)py$, 1 (**Figure 7**) . Using the parallel version of GAMESS[26] we can model the parent Cp_2ME, 2, in conjunction with these experimental studies. Calculated ME bond lengths (in Å) at the RHF level are (experiment in parentheses) ZrO = 1.76 (1.804(4)), ZrS = 2.28 (2.334(2)), ZrSe = 2.42 (2.480(1)), ZrTe = 2.68 (2.729 (1)).[25a] The results are of equal quality for Hf analogues.[25b] The Ti-oxo bond length in Cp_2TiO is 1.61 , in good accord with TiO = 1.665(3) in $Cp^*_2Ti(O)(4$-phenyl-py) ($Cp^* = \eta^5$- C_5Me_5).[27] Uniformly good agreement between theory and experiment from the lightest (Cp_2TiO) to heaviest (Cp_2HfTe) member in the series is a powerful demonstration of the ability of parallel codes to open up all areas of the Periodic Table to computation.

The example discussed above highlights two important points about the promise of parallel computing. On a standard workstation, geometry optimization and calculation of the energy hessian for a complex such as 2 can take several weeks, but just a few hours to a day on a parallel platform depending on the number of processors and their power. Vast reductions in wall clock time are thus realized through the use of parallel algorithms and architectures. A second related point is that parallel supercomputers make it possible to more closely model experimental systems. Making the model as close as possible to an experimental system has important scientific implications - errors between theory and experiment can be more confidently ascribed to deficiencies in the model or deductions based on experimental evidence. Although 2 is not a perfect model of 1 it is significantly larger than is feasible to study without parallel computers; further improvements in methods and technologies will enable direct study of 1. With parallel computing more realistic model complexes with the bulky ligands that organometallic chemists use to engender kinetic and thermodynamic stability can be studied. In other words, the chemistry that occurs in the CPU more closely resembles that which occurs in the test tube. Thus, the promise of parallel computing lies in the more productive collaborations between theory and experiment it affords through the study of larger, more accurate models in a shorter period of time. This is an important consideration in meeting the grand challenges of computer-aided design in catalysis and advanced materials, where transition metals play a very important roles.

V. Graphics for the Parallel World

The detailed nature of chemical questions being asked and hence the degree of complexity in molecular blueprints are increasing at a rate only manageable by advanced

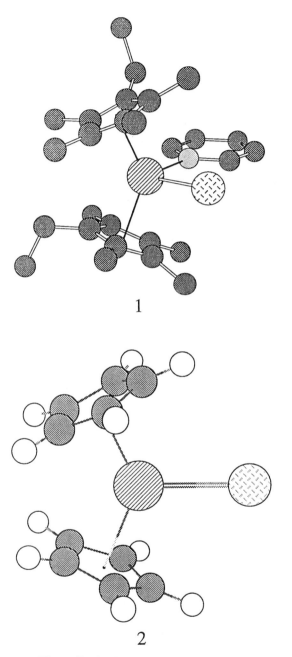

Figure 7. 1. Experimental Complex.
2. Computational Model.

computing methods, as exemplified with the applications of parallel GAMESS given above. Despite the extraordinary abilities of modern hardware technology and coding methods to manipulate the raw data, the rate limiting step in harmonizing the intricacy and precision required to push forward these chemical frontiers ultimately comes down to the process of man-machine information transfer. Along this line, words are to scalar processing what images are to parallel processing, thus, the development of versatile and facile three-dimensional visualization tools is the key to any successful human interface in this endeavor.

Quantum Mechanical View (QMView)[28] has been designed to provide the chemist with an expansive array of molecular perspectives. QMView is an integrated visualization package which capitalizes on the increased capabilities of new graphics systems to profile three-dimensional molecules not only by their common ball-and-stick or space-filling models, techniques which convey limited geometrical information, but also by molecular orbitals, electron densities (differential and absolute), electrostatic potential gradients, vibrational normal modes, regional hydro- or lipophilicity. Each profile can be adjusted, updated and presented three-dimensionally, fully colorized and in real time.

Figure 8 shows the top level interface illustrating the various capabilities of QMView with the five icons: display of 1) structure, 2) vibrational modes, 3) electron properties, 4) molecular orbitals, and 5) special features. The special features include options to run in distributed mode, and special educational tutorials. Surface data (i.e., 3 and 4) can be displayed in various manners, including options of two- or three-dimensional pictures with choices of net, solid[29] or transparent surfaces.

The second icon from the left in **Figure 8** shows a static schematic of the imaginary vibrational mode for the bowl-to-bowl interconversion of corannulene, as calculated with local density functional theory (LDA)[30]. QMView displays vibrational motions in real time after selection of the particular mode frequency from a pull-down menu. There are mouse-driven buttons to choose the number of frames for which to display the vibration, and controls over the smoothness and amplitude of the motion.

The third icon from the left in **Figure 8** illustrates the total electron density of the anthracene photodimer, lepidopterene[31], in terms of a net surface, as calculated using GAMESS. This is one of the largest geometrical optimizations performed on the SDSC Intel Paragon to date, with 494 basis functions at the 6-31G(d,p) level of theory on 32 (32 MB) nodes.

Figure 9 depicts the highest occupied, bonding molecular orbital (HOMO) superimposed on a transparent electron density surface of the C_{20} fullerene molecule, as calculated using the local density approximation [32]. Theoretical calculations have allowed researchers to asess the stability of this molecule relative to other isomers [33].

An added feature of QMView is an interface to run in a distributed mode with the parallel platforms, specifically the Intel Paragon at SDSC. In distributed mode, parallel GAMESS, running on the Intel Paragon, computes the information that QMView displays in real time. The display includes the last configuration that has been calculated by GAMESS, along with (optionally) two inset x-y graphs, one of which

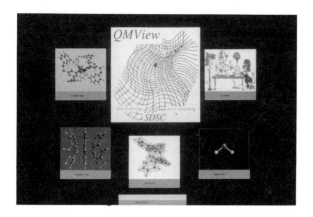

Figure 8. The first layer of the graphical user interface of QMVIEW. The icons
represent a sampling of QMVIEW functions (chosen by picking with the mouse):
Upper left: Depiction of the molecular structure of kuratowskaphane.
Lower left: Depiction of the bowl-to-bowl vibrational motion in corannulene.
Lower middle: Depiction of the total electron density of lepidopterene.
Lower right: Depiction of the molecular orbitals for water.
Upper right: This icon allows one to view a) a depiction of a calculation
as it is running on a supercomputer platform (e.g., CRAY or Paragon), and b)
various tutorials as mentioned in the text.
RESEARCH: This icon allows visualization of user-generated input.

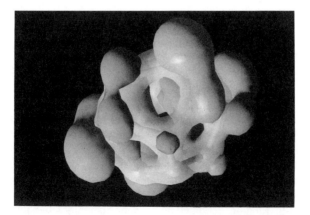

**Figure 9. Depiction of the highest occupied, bonding molecular orbital, superimposed
on an electron density surface of the C_{20} fullerene molecule.**
Calculations were performed using the Local Density Approximation.

monitors the energetics as a function of the algorithmic iteration, and the other which monitors the root-mean-square of the minimization procedure with respect to geometry as a function of the algorithmic iteration. This feature allows one to monitor a long quantum chemical calculation to make sure it is proceeding correctly and efficiently.

Acknowledgments

Studies of transition metal multiple bonding at Memphis are supported in part by grants from the Air Force Office of Scientific Research, the Petroleum Research Fund, and the National Science Foundation (CHE-9314732). The work at Iowa State and Phillips Laboratory was supported by grants from the Air Force Office of Scientific Research, the National Science Foundation, and ARPA. Communication of results prior to publication and helpful discussions with Prof. Parkin (Columbia) are gratefully acknowledged. Parallel computers at CalTech, the Cornell National Supercomputer Facility, the San Diego Supercomputer Center, Kirtland Air Force Base, and the Army High Performance Computer Center at the University of Minnesota were used for carrying out this research and all are gratefully acknowledged.

Literature Cited

1. Windus, T.L.; Schmidt, M.W.; Gordon, M.S. *ACS Symposium Series*, preceeding paper.

2. Matsunaga, N.; Gordon, M.S.; *J. Am. Chem. Soc.* submitted.

3. Boatz, J.A.; Gordon, M.S. *J. Phys. Chem.*, **1988,** *92*, 3037.

4. L'Esperance, R.P.; West, A.P.; Van Engen, D.; Pascal, R.A. *J. Am. Chem. Soc.*, **1991,** *113*, 2672, and references contained therein.

5. Kwochka, W.R.; Damrauer, R.; Schmidt, M.W.; Gordon, M.S. *Organomet.*, in press.

6. Forgacs, G.; Kolonits, M.; Hargittai, I. *Struct. Chem.* **1990,** *1*, 245.

7. Shen, Q.; Hilderbrandt, R.L. *J. Mol. Struct.* **1980,** *64*, 257.

8. Pestunovich, V.A.; Shterenberg, B.Z.; Lippma, E.T.; Myagi, M.Ya.; Alla, M.A.; Tandura, S.N.; Baryshok, V.P.; Petukhov, L.P.; Voronkov, M.G. *Doklady Phys. Chem. (English Translation)* **1977,** *258*, 587.

9. Schmidt, M.W.; Windus, T.L.; Gordon, M.S. *J. Am. Chem. Soc.* submitted

10. Dewar, M.J.S.; Zoebisch, E.G.; Healy, E.G.; Stewart, J.J.P. *J. Am. Chem. Soc.* **1985,** *107*, 3902.

11. Stewart, J.J.P. *J. Comp. Chem.* **1989,** *10*, 209, 221.

12. (a) Kirkwood, J.G *J. Chem. Phys.* **1934,** *2*, 351; (b) Onsager, L. *J. Am. Chem. Soc.* **1936,** *58*, 1486; (c) Szefan, M.; Karelson, M.M.; Katritzky, A.R.; Koput, J.; Zerner, M.C. *J. Comput. Chem.* **1993,** *14*, 371.

13. Boys, S.F. in *Quantum Science of Atoms, Molecules, and Solids;* Lowdin, P.-O., Ed.; Academic Press, NY, **1966**

14. Haaland, A. *Angew. Chem. Int. Eng, Ed.* **1989,** *28*, 992.

15. Windus, T.L.; Schmidt, M.W.; Gordon, M.S. *J. Am. Chem. Soc.* submitted.

16. Bonaccorsi, R.; Palla, P.; Tomasi, J.; *J. Am. Chem. Soc.* **1984**, 106, 1945.

17. Hehre, W.J.; Ditchfield, R.; Pople, J.A. *J. Chem. Phys.* **1972**, *56*, 2257.

18. Dunning, T.H., Jr; Hay, P.J. in *Methods of Electronic Structure Theory,* Schaefer, H.F. III, Ed. Plenum Press, NY, **1977**, 1.

19. Krishnan, R; Pople, J.A. *Int. J. Quantum Chem.* **1978**, *14*, 91.

20. Dunning, Jr., T.H. *J. Chem. Phys.* **1989**, *90*, 1007.

21. Ding, Y.; Krogh-Jespersen, K. *Chem. Phys. Lett.* **1992**, *199*, 261.

22. Cotton, F. A.; Wilkinson, G. *"Advanced Inorganic Chemistry;"* 5th Ed.Wiley: New York, **1988.**

23. Cundari, T. R.; Gordon, M. S. *Coord. Chem. Rev.* - submitted.

24. Zerner, M. C.; Salahub, D. "The Challenge of d- and f-Electrons;" ACS:Washington, D. C., **1989.**

25. TM=Chalcogen complexes: (a) Parkin, G.; Howard, W. A. *J. Am. Chem. Soc.* **1994**, *116*, 606; (b) Parkin, G.; Howard, W. A. *J. Organomet. Chem.* - in press; (c) Christou, V.; Arnold, J. *J. Am. Chem. Soc.* **1992**, *114*, 6240; (d) Diemann, E.; Mller, A. *Coord. Chem. Rev.* **1973**, *10*, 79. TM=Phosphinidene complexes: (e) Cowley, A. H.; Barron, A. R. *Acc. Chem. Res.* **1988**, *21*, 81; (f) Hitchcock, P. B.; Lappert, M. F.; Leung, W. P. *J. Chem. Soc., Chem. Comm.***1987**, 1282; (g) Schrock, R. R.; Cummins, C. C.; Davis, W. M. *Angew. Chem., Int. Ed. Engl.***1993**, *32*, 756; (h) Stephan, D. W. ; Hou, Z.; Breen, T. C. *Organometallics* **1993**, *12*, 3158. TM=Tetralide complexes: (i) Petz, W. *Chem. Rev.* **1986**, *86*, 1019; (j) Herrmann, W. A. *Angew. Chem., Int. Ed. Engl.* **1986**, *25*, 56.

26. Schmidt, M. W.; Baldridge, K. K.; Boatz, J. A.; Jensen, J. H.; Koseki, S.; Matsunaga, N.; Gordon, M. S.; Nguyen, K. A.; Su, S. Windus, T. L.; Elbert, S. T.; Montgomery, J.; Dupuis, M. *J. Comp. Chem.* **1993**, *14*, 1347.

27. Smith, M. R.; Matsunaga, P. T.; Andersen, R. A. *J. Am. Chem. Soc.* **1993**, *115*, 7049.

28. Baldridge, K.K.; Greenberg, J.P. *J. Mol. Graphics*, **1994**, submitted.

29. Lorensen, W.E. *Computer Graphics*, **1987**, *21*, 163.

30. Borchardt, A.; Baldridge, K.K., Fuchicello, A.; Kilway, K.; Siegel, J.S. *J. Am. Chem. Soc.* **1992**, *114*, 1921.

31. VernonClark, R.; Battersby, T.; Gantzel, P.; Chadha, R.; Baldridge, K.K.; Siegel, J.S. *J. Am. Chem. Soc.*, **1994**, submitted.

32. Kawai, R. Unpublished software employing the Kohn-Sham density functional theory: Hohenberg, P.; Kohn, W. *Phys. Rev. B* **1964,** *136*, 864; Kohn, W.; Sham, L.J. *Phys. Rev. A* **1965**, *140*, 1133.

33. Taylor, P.R.; Bylaska, E.; Weare, J.H.; Kawai, R. *Phys. Rev. Lett.*, **1994**, submitted.

RECEIVED January 10, 1995

Chapter 4

Object-Oriented Implementation of Parallel Ab Initio Programs

C. L. Janssen, E. T. Seidl, and M. E. Colvin

Sandia National Laboratories, Mail Stop 9214, Livermore, CA 94551

Efficient implementation of *ab initio* methods on advanced computer architectures requires rethinking the algorithms and coding practices currently in use. This creates an opportunity to experiment with new software development methodologies while building the next generation of codes. We have chosen an object oriented approach using the C++ programming language. Our goal is a production-quality set of computational chemistry programs that run efficiently on scalar, distributed, shared memory, and massively parallel computers. We will describe our massively parallel quantum chemistry program with emphasis on understanding how well the object oriented approach facilitates the development of scientific software. We will also examine the effects that our design choice has had on efficiency, code reuse, and complexity.

Rapid advances in computer hardware and software technology have made the implementation of efficient *ab initio* quantum chemistry programs a continuing effort. Since these methods are so computationally intensive, programmers must redesign code to take maximal advantage of the currently available computer hardware. For example, programs to compute the two electron integrals were originally constrained to fit into a few kilobytes of memory. They have now been replaced by much larger programs that take advantage of vectorized computer architectures. This constant algorithmic evolution has not been matched by changes in software development methods. After early machine language versions of quantum chemistry programs, the community quickly settled on the FORTRAN programming language, which has seen few changes in the last two decades.

There are several reasons for the slow change of implementation language; one of the most important has been that, until recently, compilers of other languages could not generate executable programs as efficient as the best FORTRAN compilers. Although

0097–6156/95/0592–0047$12.00/0

high performance compilers are available in a number of new languages suitable for scientific computing, the quantum chemistry community continues to be reluctant to use them. This continued reluctance is due in part to the time required to learn new computer languages and, in our opinion, an underestimation of the value of modern programming methods. In particular, object oriented programming methods and languages provide powerful ways of organizing programs that are particularly useful for dealing with complex quantum chemistry programs, without sacrificing efficiency. Moreover, these new computer languages will help reduce the burden of creating programs that are efficient on diverse computer architectures ranging from stand alone workstations to massively parallel supercomputers.

The Object Oriented Approach

One of the main advantages of object oriented programming is that it provides mechanisms for hiding the complexity of large software systems. There are several ways in which the object oriented approach helps in this regard, but we will focus on the ways in which the object oriented approach eases reuse of code, improves portability, and produces more reliable programs. These benefits are provided by object oriented languages through two primary mechanisms: encapsulation and abstraction (and concomitant specialization).

Encapsulation. Encapsulation is the grouping of related data and the operations that manipulate these data together into a new data type, sometimes called a "class". The concept of encapsulation is familiar to many users of traditional programming languages, such as C, that allow the programmer to define new data structures. Object oriented languages take this one step further by giving the programmer of this new data type control over how the data that are encapsulated within the new type can be accessed and manipulated. The programmer decides whether or not individual pieces of data that compose the new type will be accessible to other users of the class. Additionally, the programmer writes subroutines that are considered a part of the new data type and that have privileged access to all of the component data of that type.

The component data and functions belonging to a data type are known as its "members". Those members accessible to all users of the class are collectively referred to as the interface. The key to successful object oriented design is the appropriate choice of the interface, because as long as it is not necessary to change the interface to adapt the data type to a new situation, it is not necessary for any other users to be aware of the internal workings of the the data type.

Just as an integer data type in a conventional programming language has specific instantiations while a program is running, for example, the integer variable nbasis may have the value 731, an object oriented class will have instantiations in a running program. These instantiations are the "objects" in object oriented programming.

There is an important difference between member functions and ordinary functions. Ordinary functions typically operate on external data that are passed into the function. Member functions are associated with a specific object and usually act to modify or extract data from this object. For example, a matrix object might have a matrix inversion member function which would calculate and return the inverse of the matrix held by the object.

For a more detailed example, let us consider how a simplified self-consistent field (SCF) wavefunction might be encapsulated in the SCFWAVEFUNCTION data type (see Table I). The members of SCFWAVEFUNCTION are separated into two categories. The

Table I: The SCFWAVEFUNCTION data type.

Private Member Data:		
Type	Name	Description
MOLECULE	molecule	The nuclei and their positions.
REAL	E	The energy.
MATRIX	coef	The molecular orbital coefficients.
BOOLEAN	current	True if E and coef are current.
Public Member Functions:		
Type	Name	Description
REAL	energy()	Returns the energy of the molecule.
MATRIX	coefficients()	Returns a copy of coef.
—	geom(MATRIX)	Sets the geometry of molecule to the given MATRIX object.

"private" member data are those data that users of SCFWAVEFUNCTION objects cannot access directly. They are considered internal to SCFWAVEFUNCTION and they can only be accessed by the designer of this class within the code for the member functions. Users of this class have access to the public member functions only. That is, they can only retrieve the energy or SCF coefficients, or change the molecular geometry.

This greatly improves the integrity of the code, since it defines the only mechanisms by which the user can modify the object. If the user wants to change the geometry, then this can only be done through the geom(MATRIX) member. If the geometry is changed, the member current would be set to false. The next time energy() is called, it would check current, find that it was false, recompute the energy, and finally return the answer. Subsequent calls to energy() would find that the stored energy was current and would return the energy without duplicating the computation. Now it is no longer up to the user to remember to update the energy when the geometry is changed. This is guaranteed by the SCFWAVEFUNCTION class itself.

Abstraction. It is frequently useful to design generic data types for which a variety of specific implementations are desired. (These are called abstract data types.) While these data types specify a complete interface, implementations of some of their member functions are deferred. New data types can inherit the interface of an abstract data type and implement the deferred member functions. (The new type is called a specialization of the abstract data type.) With this approach, two desirable features are obtained. First, a foundation is provided on which new data types can be based. This lets the new data types reuse the pieces of code which could be implemented for the more general data type. Furthermore, abstraction allows a piece of code to be written that uses only the interface of the abstract data type. This means that any data type which has that abstract data type as its foundation can be given to that piece of code.

For example, in a parallel quantum chemistry program the data type MATRIX may exist in several different forms. It could be distributed across or replicated on the nodes of the system or, on some architectures, it could reside in shared memory. Although the underlying storage scheme for MATRIX is unspecified, the concept of a matrix is quite well defined. It should support matrix multiplication, diagonalization, and all of the other operations associated with matrices. Any user that used only these general matrix operations and was not concerned about the internal details of the matrix, such as whether or not it was distributed, could write code solely in terms of MATRIX which would not have to be reimplemented or even recompiled for each of the specialty matrices.

Going back to the example of SCFWAVEFUNCTION, we find that abstraction could be useful here as well. Table II shows the members of the WAVEFUNCTION class. (Note

Table II: The WAVEFUNCTION data type.

Protected Member Data:		
Type	Name	Description
MOLECULE	molecule	The nuclei and their positions.
REAL	E	The energy.
BOOLEAN	current	True if E and coef are current.
Protected Member Functions:		
Type	Name	Description
—	update()	Recomputes the energy. This function is deferred, because WAVEFUNCTION doesn't know how to compute the energy.
Public Member Functions:		
Type	Name	Description
REAL	energy()	Returns the energy of the molecule.
—	geom(MATRIX)	Set the geometry of molecule to the given MATRIX object.

that we have introduced a new access type in addition to "private" and "public". "Protected" members can be used by classes which inherit from the class with protected members, but they are not accessible to other users of the class.)

Specialization is the process of providing an implementation for an abstract data type. This is done by defining a data type that inherits the properties of an abstract data type and implements all of the member functions deferred by the abstract data type. Table III shows the SCF specialization of the WAVEFUNCTION data type. Since SCF inherits the members of its base class, WAVEFUNCTION, it ends up with the same interface as the original SCFWAVEFUNCTION data type. Only now, part of its code has already been implemented by its base class. Furthermore, any code which must com-

Table III: The SCF data type.

Private Member Data:		
Type	Name	Description
MATRIX	coef	The molecular orbital coefficients.
Private Member Functions:		
Type	Name	Description
—	update()	Recomputes the energy.
Public Member Functions:		
Type	Name	Description
MATRIX	coefficients()	Return a copy of coef.

pute a molecular energy can be passed any WAVEFUNCTION class whether it be an SCF or an MP2 or any other class based on WAVEFUNCTION.

The relationships between the classes are shown in Figure 1. This shows classes in

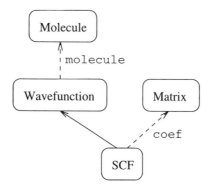

Figure 1: The SCF class hierarchy.

boxes and inheritance relationships as solid lines pointing from the specialized class to the abstract base class. The dashed lines represent containment, that is, data membership, and point to the contained data type. These lines are labeled with the name of the member.

We can benefit from applying the process of abstraction and specialization at a higher level. Suppose we wanted to optimize the energy associated with our wavefunction with respect to changes in the nuclear coordinates. The optimization package does not need to know whether or not we are computing the energy of a molecule or that the coordinates are nuclear coordinates. It only needs a function which, given a set of parameters, can compute the function's value and possibly its gradient. By basing the WAVEFUNCTION class on a FUNCTION class, we can make the optimization routines more general and reusable.

Object Oriented Languages

Object oriented programming techniques can be applied in nearly any language (*1*), but their use is greatly facilitated by the use of object oriented programming languages, and we will therefore restrict our attention to the use of object oriented languages, such as C++ (*2*).

However, since performance is critical, we compared the efficiency of C++ with other programming languages traditionally used for scientific applications. We implemented a loop-unrolled, double precision, matrix multiply routine in FORTRAN, C, and C++. We compiled these test programs on an SGI Onyx workstation (150 Mhz R4400 CPU, 26 MFLOPS 100 × 100 Linpack (*3*)) using the SGI FORTRAN, C, and C++ compilers. As shown in Table IV, all three produced nearly equal performance for 100 × 100 matrix multiplies—in the range 36-40 MFLOPS. Needless to say, carelessly using member functions can lead to dramatic reductions in the performance of C++ (as could overzealous use of function subroutines in FORTRAN). To demonstrate this, the final row in the table is for a C++ matrix multiply in which a member function is invoked to access each matrix element, which reduces the performance to 3 MFLOPS. Clearly, excessive use of member function calls must be avoided and this is done in object oriented languages by carefully choosing the interface. In this case, the matrix interface has a member that is able to efficiently multiply two matrices.

Table IV: Speed of a matrix multiply written in several languages.

Language	Rate (MFLOPS)
FORTRAN	36·4
C	39·7
C++	39·5
C++[a]	3·1

[a] A function call was used for each element access.

The object oriented languages can be broadly grouped into weakly and strongly typed languages. The strongly typed languages are similar to FORTRAN and C, in which each datum is associated with a particular type such as integer or float, while weakly typed languages do not permit types to be associated with symbols. Typically, compilers for strongly typed languages have the advantage that they can ensure that only appropriate operations are performed on a given datum and they can more easily optimize the generated code. On the other hand weakly typed languages are more flexible. For quantum chemistry, which places a premium on execution speed, the strongly typed languages are the logical choice, at least for the compute intensive kernels. For this reason we will focus exclusively on the strongly typed languages.

Several strongly typed object oriented languages exist, of which the most widely used is C++. Commercial as well as freely distributed C++ compilers are available for nearly all architectures, including massively parallel machines. This is one of the main reasons we have chosen C++ for the implementation of our massively parallel quantum chemistry program. Nevertheless C++ is not a perfect object oriented language; it provides only the basic machinery that permits object orientation. Other languages typically layer upon this foundation a set of commonly needed data types, such as simple arrays, sets, strings, and even complex data types such as those needed to build a graphical user interface to an application. Furthermore, it is common for object oriented languages to provide built-in facilities such as a mechanism to automatically reclaim memory as soon as it is no longer needed. Also, methods that retrieve "metainformation" about a data type, such as its type name and relationships to other types, are provided by some languages. Another important facility that is frequently provided is persistence, that is, the ability to save an object to a disk or other device so that it can be restored by another program or perhaps moved to another processor. None of these data types or facilities are provided by C++. It does, however, provide the framework for the programmer to build these data types and facilities, but this task must be undertaken by the applications developer. This lack of basic support by C++ can seriously jeopardize code reuse since individual developers may use different approaches to provide the functionality missing in C++. (Consult the appendix for a discussion of how we have implemented some of the facilities missing in C++.)

These drawbacks, as well as other, more technical, problems with C++ raise the question of whether it is worthwhile learning a complex new language. However, in our case, implementing quantum chemistry codes on massively parallel machines and other parallel architectures requires a substantial rewrite of the codes and the object oriented approach is a particularly sensible way to deal with the complexity of developing software that is efficient on several different computer architectures. Furthermore, although C++ currently adds unnecessary complexity to object oriented programming, this situation is likely to be ameliorated in the future. New languages will come along, or other languages will become better accepted, or C++ itself will evolve to remedy its problems. Whatever language we program in ten years from now, it will very likely have object oriented features and will support the same abstractions we are developing now. Ultimately, it is the proper choice of abstractions, and not the choice of language, that is the key to successful object oriented design.

The primary document for the C++ standard (2) can be consulted for more details.

Applications to *Ab Initio* Chemistry

We are in the process of applying object oriented design principles to our massively parallel quantum chemistry (MPQC) codes which can perform low order *ab initio* computations on medium sized biochemicals and portions of macromolecules with up to hundreds of atoms. MPQC can currently be used to compute SCF energies and gradients as well as second order Møller Plesset (MP2) (4) and open-shell (OPT2) (5) perturbation theory energies. Efficient internal coordinate optimization methods allow for the rapid determination of molecular geometries at the SCF level of theory. Since cutoffs are used to drastically reduce the number of integrals needed for the large systems we study, we have taken care to parallelize all of the $O(N_{basis}^3)$ steps (matrix multiplication, diagonalization, orthogonalization, etc.) to prevent these steps from becoming bottlenecks.

Furthermore, N_{basis}^2 may be large compared to the amount of memory available on each processor, so we have the ability to distribute matrices among the nodes as described in (6).

MPQC running on an Intel Paragon is being routinely used for chemical studies. Table V illustrates typical calculations and timings for a single SCF gradient calculation.

Table V: Timings for SCF gradient calculations running on an Intel Paragon.

Molecule	Point Group	N_{basis}	$N_{processor}$	Time (hours)
Methyl-α-cellobiose	C_1	480	120	2·83
Methyl-α-cellobiose	C_1	480	240	1·50
Methyl-α-cellobiose	C_1	480	480	0·83
Porphyrin	S_4	420	256	0·34
Acetylaminofluorene	C_1	294	256	0·55
Phthalocyanine	D_{2h}	970	512	1·05

Object Oriented Design of MPQC. A simplified view of a portion of the MPQC inheritance hierarchy is shown in Figure 2. This is a more realistic reworking of the SCF hi-

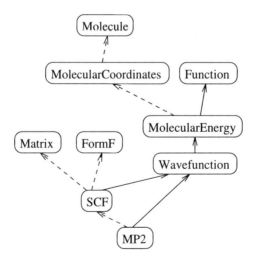

Figure 2: The MPQC inheritance hierarchy.

erarchy illustrated in Figure 1. Features include the use of the FUNCTION abstract class which maps a set of coordinates to the function's value (and possibly its gradient). A separate set of optimization classes can make use of any specialization of FUNCTION. The MOLECULARENERGY class is a specialization of FUNCTION (it happens to be an abstract specialization) which interprets FUNCTION's coordinates as the specification of a nuclear geometry and FUNCTION's value as the energy of the molecule. Since there are many ways to specify a geometry and the actual computation of the energy is usually done in terms of Cartesian coordinates, MOLECULARENERGY contains a class, MOLECULARCOORDINATES, which is used to convert between the coordinates that are needed by FUNCTION and the Cartesian coordinates. Note that at this level, the MOLECULARENERGY class is completely generic to any method of calculating the molecular energies. For the case of *ab initio* methods, it is useful to create specialization of the MOLECULARENERGY class, WAVEFUNCTION, which adds members to compute electron densities and wavefunction values at points in space. Finally, WAVEFUNCTION is specialized into a fully implemented class for each *ab initio* method such as SCF or MP2. The SCF specialization of WAVEFUNCTION contains objects such as the SCF coefficients (which are of type MATRIX) and an object to form the Fock matrix (which is of type FORMF). The WAVEFUNCTION class has also been specialized to the MP2 class in Figure 2 which includes an SCF object to store the reference wavefunction.

An SCF object can be used to compute an energy or gradient. Since the SCF class is a specialization (via WAVEFUNCTION) of the FUNCTION class, it is guaranteed to be compatible with the OPTIMIZE class, shown in Figure 3, which can be used to optimize the SCF geometry. In practice, we use the QUASINEWTON specialization of OPTIMIZE

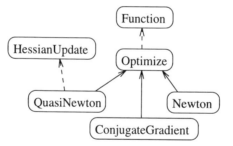

Figure 3: The OPTIMIZE inheritance hierarchy.

which would contain an SCF specialization of FUNCTION.

Nothing specific to parallelism appears in Figures 2 and 3. This is precisely the goal of the object oriented approach; we want to hide the complexity of parallelism as much as possible. The core types of MPQC provide an adequate foundation for all computer architectures. To understand how parallelism fits into this scheme, we will look at the MATRIX, FORMF, and MP2 classes in more detail.

The MATRIX Class. Matrix operations are among the more common tasks of computational significance that a quantum chemistry application must perform. Thus it is desirable to develop highly optimized matrix classes for each computer architecture. In the object oriented approach, this is accomplished by defining an abstract class and a specialization of this class for each computer architecture, as shown in Figure 4. Unfor-

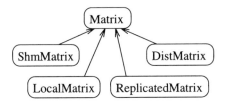

Figure 4: The MATRIX inheritance hierarchy.

tunately, the existing object oriented matrix libraries are optimized for particular computer architectures. Our goal is to design a set of matrix operations for the abstract MATRIX class that is very general and will satisfy our needs for all architectures. We plan to support simple matrices that provide efficient matrix computations for uniprocessors, shared memory machines, clusters of processors with enough memory to store all matrices connected by a relatively slow network (workstations on an LAN), clusters of processors with enough memory to store all matrices and a fast interconnect network (a massively parallel machine such as the Intel Paragon), and clusters of machines that do not have enough memory to hold entire matrices (only a fast interconnect network would work well in this case).

The general operations mentioned above define the interface of the abstract MATRIX class, a portion of which is outlined in Table VI. This table omits the standard linear algebra routines, such a matrix inversion, multiply, etc. that are also part of the

Table VI: A portion of the MATRIX interface.

Public Member Functions:		
Type	Name	Description
MATRIX	copy()	Return a copy of this matrix.
—	assign(REAL)	Assign all elements to the given number.
—	assign(MATRIX)	Assign to this matrix the given matrix.
—	accum(MATRIX)	Add to this matrix the given matrix.
MATRIX	sum(MATRIX)	Return the sum of this and another matrix.
—	set(INTEGER, INTEGER, REAL)	Set an element of this matrix.
REAL	get(INTEGER, INTEGER)	Return an element of this matrix.
—	element_op(ELEMENTOP)	Perform the given operation on all elements of this matrix.

interface as well as other utility methods that implement automated memory management, persistence, and dynamic typing.

Although the MATRIX class is abstract, it can implement some of these members without limiting its generality. For example, the sum() member is implemented in MATRIX using the copy() and accum() members. The implementations of copy() and accum() are deferred to the specializations of MATRIX, since they require specific information about how the matrix elements are stored. An alternate choice would have been to implement accum() in terms of and sum() and assign(); however, this would result in the unnecessary creation of a temporary matrix. (Such considerations of storage and performance requirements are essential to developing classes for high performance applications.) Ideally, object oriented matrix classes should permit use of convenient member functions, such as sum(), while allowing access to more efficient members, such as accum().

A more detailed example of some of the trade-offs between efficiency and generality arises when deciding how to provide the user access to the elements of the MATRIX objects. Since some specializations of MATRIX distribute elements among the processors, individual access to each element could be slow. Hence, the MATRIX member functions get() and set() listed in Table VI will not alone be adequate. (We have included these functions since there are many cases where efficiency is not important and omission of get and set would make the matrix package less flexible.) An alternative strategy to retrieving the individual matrix elements before an operation is to distribute the operation request to wherever the matrix elements are located. To this end another method, element_op(), has been added to the MATRIX class. The element_op() member takes as an argument an ELEMENTOP object, which is capable of processing each of the elements in the matrix in an efficient manner.

The use of the element_op() member is illustrated in Figures 5 and 6 for the computation of the overlap matrix on three processors (one of which is the "host" that manages the parallel computation). In this example, the overlap matrix, S, is of type

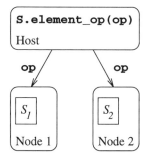

Figure 5: The first step in the overlap matrix computation.

DISTRIBUTEDMATRIX and a piece of it, S_1, resides on node 1 and the rest, S_2, is stored on node 2. In the first step, shown in Figure 5, the host process creates the op object. The op object is of type OVERLAP which is a specialization of ELEMENTOP. This op

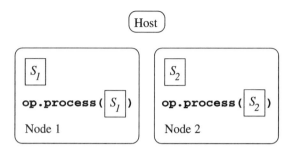

Figure 6: The second step in the overlap matrix computation.

is then given as an argument to the `element_op()` member. In the next step, shown in Figure 6, the `op` is broadcast to the nodes, using the persistence mechanism to duplicate this object. Upon receiving `op`, each node passes its locally held pieces of the overlap matrix to `op`, which fills in the values. The matrix classes allow the granularity of the pieces of the matrices to be chosen such that they make the procedure efficient, which in this case means that all basis functions in a shell must be grouped together.

The FORMF Class. Although most quantum chemical algorithms are formulated in terms of matrix equations, more specialized operations are required for their efficient computation, especially on multiprocessor computers. One example is the parallelized Fock matrix formation, where the two electron integrals and perhaps the F, H, and P matrices are distributed.

$$F_{pq} = H_{core} + \sum_{rs}^{n} P_{rs}(2(pq|rs) - (pr|qs))$$

This equation can be implemented in two ways; abstract operations could be developed for the MATRIX class that can do this sort of contraction for the general case. A more efficient alternative is a class written to implement the Fock matrix formation, FORMF, which is encapsulated within the SCF class (Figure 2). This class has specializations optimized for particular architectures, as shown in Figure 7. The latter approach is simpler and is currently being used in MPQC. Since the SCF object must first determine whether a local or distributed MATRIX is actually being used for the density, some mechanism is needed to identify the matrix specialization (which may not be known until run time). Our dynamic typing system makes this possible with C++ (see the appendix); other object oriented languages support this directly. The appropriate specialization of FORMF is created which knows the specific form of the matrix being used and thus it can access the data in the most efficient way. This approach is an example where a localized breakdown of abstraction is required to yield optimal efficiency from the machine.

The MP2 Class. To achieve the maximum efficiency, we must accept the fact that certain wavefunctions may have very different implementations for different architectures.

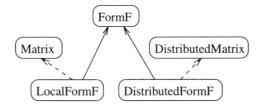

Figure 7: The FORMF inheritance hierarchy.

This is the case with our MP2 routines, because the entire algorithm has been reorganized to minimize memory use and communications. In this case the inheritance hierarchy includes a completely separate specialization of WAVEFUNCTION to support MP2 on parallel architectures. (Figure 8.)

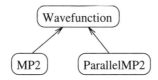

Figure 8: The MP2 inheritance hierarchy.

Conclusions

The object oriented approach can be used to manage the complexity of large software systems. Object oriented languages allow programmers to encapsulate, abstract, and specialize data types. These facilities allow code to be written that is easier to reuse and more general. When it is not possible to obtain satisfactory performance out of general code, object oriented programming methodology allows programmers to isolate necessary machine specific code as much as possible from the rest of the application. These features make object oriented programming a particularly good way to implement complex parallel quantum chemistry programs.

A continuing hindrance to the field of *ab initio* quantum chemistry is that nearly all of the software infrastructure is reimplemented in each new quantum chemistry project. As a result, there are fundamental incompatibilities in the quantum chemistry software developed by different research groups. A central goal of this research is to determine the feasibility of a set of object oriented building blocks for quantum chemical software. If such object libraries can be made sufficiently versatile and efficient on a wide range of computer architectures, they should point the way towards a different future for quantum chemical methods development. New theoretical methods could be rapidly assembled from available classes and new developers could fully take advantage of an existing legacy of quantum chemical software.

Appendix

As previously stated, the MPQC program is implemented in C++, which omits several useful programming features that are common in other object oriented programming languages. Fortunately, C++ is flexible enough to allow inclusion of most of these features with a modest amount of additional programming. Some of the missing features that we have implemented for use in MPQC are automated memory management, dynamic typing, and persistence. These utility classes are available from the authors (email janssen@netcom.com).

Memory Management. C++ was designed as a superset of its predecessor, C, and uses C's memory management mechanism. This involves explicit function calls to allocate and release memory. It is frequently the case that a single piece of memory is shared by many objects, to avoid allocating extra storage and unnecessarily copying the data. However, it then becomes difficult to determine when the memory is no longer in use and should be released to the system so it can be reused. Releasing the memory too soon typically results in difficult-to-find errors in the program; not releasing memory at all wastes system resources. Several other programming languages use "garbage collection" techniques to reclaim unused memory without programmer intervention.

Our approach to memory management in C++ allows the programmer to choose between the standard C-style memory allocation techniques and a "reference counting" mechanism for each object created. In the latter method an integer is stored with the object that keeps track of how many references that there are to an object. This count is maintained by a "smart pointer" to the object which is implemented as a class that encapsulates a simple C-style pointer to the object. The smart pointer can be used just like a simple pointer. When an operation that eliminates a reference to an object is executed, the smart pointer will detect this and decrement the reference count. When the count becomes zero, the object is released. When a new reference to an object is created by a smart pointer the reference count is incremented. Smart pointers can be used in place of simple pointers without significant performance impact for big objects such as matrices. For small data, such as the elements of a matrix, the standard C-style pointers are used.

Dynamic Typing. Consider the accum(MATRIX) member of the MATRIX class (Table VI). This is a deferred member function so it is implemented in a specialization of MATRIX. If the accum(MATRIX) member of an DISTMATRIX (Figure 4) object is called, then it will be passed an argument of type MATRIX. This means it could only use the functions provided in the interface of MATRIX to perform the accumulation. This is not very efficient, but to improve the performance the accum(MATRIX) member function of DISTMATRIX would need to know which particular MATRIX specialization it has been passed.

Dynamic typing makes it possible to determine the specialized type of an object for which only the abstract type is known. Thus, the accum(MATRIX) member function could find out if the argument was actually a DISTMATRIX object. It could then do the accumulation in the most efficient way possible.

Persistence. Persistence is broadly interpreted to mean capabilities ranging from saving objects to a disk so that they can be reconstructed in the future to sending objects across a network so they can reside on a different node. Both capabilities are very useful in a parallel quantum chemistry code.

We have implemented a persistence mechanism that is fully integrated with the object oriented approach. All classes with persistence inherit from the SAVABLESTATE class which provides a deferred member function, `save_state`(STATEOUT), that is called with a argument that is of type STATEOUT. Each specialization of SAVABLE- STATE implements `save_state`(STATEOUT) so that it gives to the STATEOUT object each piece of data that is needed to reconstruct the object being saved. Different specializations of STATEOUT object are used to dispose of the data in different ways. For example, the STATEOUTBINFILE specialization writes each piece of data to a binary file in an architecture independent format.

Reconstruction of the object proceeds by giving an object's constructor (a special member function that initializes a new object) an object of type STATEIN. For each STATEOUT there is a corresponding STATEIN specialization that can restore data saved by the STATEOUT specialization. Because only the abstract type of member objects might be known, it is necessary to store the specialization of the object with the object so that the correct specialization can be restored. The dynamic typing mechanism provides this capacity. After the restoration is complete, we have an object that has exactly the same properties as the object that existed at the time its `save_state`() member was called.

Acknowledgments

This work was carried out at Sandia National Laboratories under contract from the U.S. Department of Energy and supported by its Division of Basic Energy Sciences. Parallel computing resources were obtained in part from the Massively Parallel Computing Research Laboratory at Sandia National Laboratories.

Literature Cited

1. Holub, A. I. *C + C++*. McGraw Hill, 1992.

2. Ellis, M. A.; Stroustrup B. *The Annotated C++ Reference Manual*. Addison-Wesley, 1990.

3. Dongarra, J.J. *Performance of Various Computers Using Standard Linear Equations Software*. Tech. Report CS-89-85, Oak Ridge National Laboratory, Oak Ridge, TN, 1994.

4. Møller, C.; Plesset M. S. *Phys. Rev.* **1934**, 46, p 618.

5. Murray, C.; Davidson, E. R. *Chem. Phys. Letters* **1991**, 187, pp 451–454.

6. Colvin, M. E.; Janssen, C. L.; Whiteside R. A.; Tong C. H. *Theor. Chim. Acta* **1993**, 84, pp 301–314.

RECEIVED January 19, 1995

Chapter 5

Ab Initio Quantum Chemistry on a Workstation Cluster

David P. Turner[1], Gary W. Trucks[2], and Michael J. Frisch[2]

[1]Scientific Computing Associates, 265 Church Street,
New Haven, CT 06510-7010
[2]Lorentzian, Inc., 140 Washington Avenue, North Haven, CT 06473

Recent advances in workstation performance and parallel programming environments have produced a new computing option for ab initio electronic structure theory calculations: parallel processing on clusters of workstations. This model, if successfully applied to the most commonly-used algorithms, promises to provide a larger number of researchers quicker time to solution, the ability to study larger chemical systems, and the availability of vast amounts of cost-effective computing resources. In this chapter, we will describe our efforts parallelizing the Hartree-Fock direct SCF energy, gradient, and second derivative evaluations in the widely-used Gaussian ab initio code system, using the Linda parallel programming model. We will address the relevant issues of the cluster programming model, and will present our results from a network of six high-performance Unix workstations. We will also briefly discuss our plans for extending the parallel performance to a larger number of processes.

Numerical computation is an important component of modern quantum chemistry, both in private industry as well as in academic research. Ab initio electronic structure theory has proven to be a valuable tool for determining structures, thermochemistry, characterizing spectra, predicting reaction mechanisms and rates, and determining parameters for more approximate methods such as molecular mechanics. In particular, the Gaussian series of programs[1] has been established as the most widely used ab initio package. This is due to its constantly improving performance and its open software architecture, which allows the newest models to be easily incorporated.

In spite of recent dramatic gains in Gaussian performance, many researchers are still limited both in the size of the molecular systems they can study and in the extent of analysis possible. The typical researcher may have two computing options available, each with its own limitations. The first option is the traditional supercomputing center. While memory capacity and CPU performance are often quite dramatic, most centers are heavily subscribed. In practice, it may be impossible to get the necessary resource allocations to thoroughly study the chemical system of interest. The second option is the researcher's own workstation. Here the

0097-6156/95/0592-0062$12.00/0
© 1995 American Chemical Society

situation is reversed: CPU performance and memory capacity are much more modest, but their availability is essentially unlimited. Typically, the researcher's workstation is but one of many located on a local area network. The primary objective of this work has been to investigate the creation of a third option, a parallel version of Gaussian that can run on a network of workstations. This objective was constrained by the desire to leave intact the underlying structure (and portability) of Gaussian.

The first step of any code optimization effort, including parallelization, is to identify those operations that dominate the quantity being optimized, whether it is CPU time, memory usage, or I/O[2]. For Hartree-Fock theory applied to large chemical systems, the limiting calculation in direct self-consistent field (SCF) methods is the generation and consumption of two-electron repulsion integrals[3]. Gaussian implements this using the Prism algorithm[4,5]. Although this is the fastest known algorithm for two-electron integral evaluation, it can still comprise approximately 90% of the required CPU time for a typical single-point energy or gradient calculation. It is therefore an obvious candidate for parallelization. Another area that can use enormous amounts of CPU time is the evaluation of Hartree-Fock analytic second derivatives. In Gaussian, a large part of this work is accomplished by a package called ChewER, which implements the Head-Gordon-Pople (HGP) algorithm for Raffenetti integral combinations[6]. The primary output of both Prism and ChewER is a set of Fock matrices. Each element of each matrix is the sum of many two-electron integral terms. This summation can be used as the basis of work partitioning among multiple processes.

Just as Gaussian is a leader in the field of ab initio computations, Linda is a leader in parallel programming. It has achieved this position through its simple programming model, portability, debugging tools, and of course, performance. It provides a framework for developing and debugging a parallel application on a single processor, and then using the same source code to generate a version suitable for a multiprocessor (shared- or distributed-memory) or a network of workstations. For these reasons, Linda was chosen as the parallel programming tool for this work.

There have been several attempts to parallelize electronic structure codes in the past, some of which have yielded good parallel speedup[7-12]. However, these efforts have primarily started from academic programs and lack commercial support for end users. Many of these codes incorporate a limited range of models and do not necessarily use the most recent and effective algorithms. The work reported herein differs from these previous works in its emphasis on using two state-of-the-art commercial software packages, Gaussian and Network Linda, and in its focus on the actual results an end-user would receive from such a combination. For each test case that we benchmarked, we have presented not only speedups for the computational kernels that were parallelized, but also for the entire Gaussian calculation, which are much more relevant to a typical end-user.

The aim of this research was to exploit parallelism on small, homogeneous networks. Because of this, it was decided to base the Linda version on the existing shared-memory multiprocessor version of Gaussian. As described below, this approach results in coarse-grain parallelism, with a large initial communication cost, no communication during parallel execution, and a smaller communication requirement following the parallel section. The large granularity suggested that a network implementation was practical. The work allocation of this method provides an equal amount of work for all processes, providing good load balancing for identical processors.

In the following sections, we provide technical background, describe the parallel implementation, and present and discuss performance results. Additionally, we identify areas for possible future parallelization. Overall, our work clearly establishes the feasibility of developing a network-parallel version of Gaussian.

Background

Gaussian Architecture. The Unix version of Gaussian is actually a sequence of executable programs called *links*. The first link (L0) is responsible for reading the user's input file and determining the sequence of links necessary to produce the desired results. Each link is responsible for initiating the next link in the sequence, using the exec() system call.

There are over a dozen links which can possibly invoke Prism and/or ChewER. For this project, the following links were chosen for their frequent and CPU-intensive use of these two algorithms:

- L502   closed and open shell SCF solution;
- L703   two-electron integral first or second derivative evaluation;
- L1002   Coupled-Perturbed Hartree-Fock (CPHF) solution and contribution of coefficient derivatives to Hartree-Fock second derivatives; and
- L1110   two-electron contribution to Fock matrix derivatives with respect to nuclear coordinates.

Of the remaining links that use these two packages for integral evaluation, most consume little CPU time, or are infrequently used. A few of them were successfully parallelized to test the robustness of the method, but no performance measurements were made for them.

Shared-Memory Gaussian. The most recent public release of Gaussian (Gaussian 92) includes versions of Prism and ChewER intended for shared-memory multiprocessors[13]. Most of the logic which implements this parallelism is located in two setup routines called PrsmSu and CherSu. These routines are responsible for process creation and synchronization (only required at termination).

Process creation is implemented using the Unix fork() system call. This gives each child process private copies of all local variables, COMMON blocks, and scalar arguments. The array arguments reside in shared memory, and each child process gets free access to all the arrays. For input arrays, no special care is needed. For arrays that are modified (either scratch arrays or output arrays), the parent allocates space from its shared-memory workspace for each of its children, and initializes the arrays appropriately. It should also be noted that the parent process calls Prism (or ChewER) after creating the child processes; that is, the parent also functions as a worker, participating fully in computing the solution.

Work allocation is done in parallel but is completely deterministic. In the Prism routine PickS4, each worker determines which batch of shell quartets it will evaluate. In the current implementation, this is a completely redundant and non-trivial computation, comprising from 3% to 4% of the overall Prism execution time. A similar approach is taken by the ChewER routine ChunkR. The effect of this redundant work is clearly seen in the results presented later.

Using the shell quartets selected by PickS4 (or ChunkR), each worker computes a partial sum for every element of every Fock output matrix. After completing their calculations, the children exit, leaving their result matrices in shared memory. When the parent has finished its calculation, it calls wait() for each child. Then it sums the matrices left behind by the children into its own result matrices. These matrices are then returned to the calling Gaussian link.

The above strategy has many benefits. This single approach allows effective parallelism whether computing a single Fock matrix (e.g., during the SCF solution), or whether computing many matrices (e.g., during the CPHF phase of the Hartree-Fock second derivative evaluation, where a separate matrix for each perturbation must be constructed). While this method requires more memory than a sub-blocking technique, it avoids the redundant integral evaluation or extra communication (and therefore synchronization) that would be required in a sub-blocking scheme (due to the random nature of a quartet's contribution to a Fock matrix).

Linda Fundamentals. Because Linda has been extensively discussed in many previous publications[14-20], only its relevant features are presented here. Linda is based on a communications abstraction called *tuple space* (TS). To the application program, TS appears as a content-addressable virtual shared memory containing data objects known as *tuples*. A tuple is a collection of typed fields; each field can be a literal value or the name of a variable. Tuples are deposited into TS with the `out` operation, and are withdrawn or read with either the `in` or the `rd` operation. These two input operations differ in that `in` removes the tuple from TS, whereas `rd` returns a copy of the tuple, leaving the original in TS. Both `in` and `rd` are blocking; that is, if the desired tuple is not in TS, the requesting process will be suspended until an appropriate tuple becomes available. The final operation of interest is `eval`. While `eval` is formally defined in terms of TS operations, for our purpose it may be thought of as creating a concurrently executing process. An `eval` requires the name of a subroutine, to be used as the entry point of the new process, and a short list of simple arguments to be provided to that subroutine.

The Linda model is independent of the underlying hardware. In order to run in parallel on a network of Unix workstations, the Linda utility `ntsnet` is used. This program is responsible for scheduling and initiating the Linda processes on the desired remote nodes. Its single required argument is the name of the parent process; all other necessary information may be provided through configuration files.

Shared Memory vs. Linda. The fundamental difference between the current Gaussian parallel model and the Linda model is the difference in semantics between `fork()` and `eval`. The processes created by `fork()` inherit all current state information from the parent, i.e., they are copies of the parent as it existed at the time of the `fork()`. In contrast, the processes created by `eval` are copies of the parent as it initially existed. This semantic difference is required in order to support a common programming environment across shared- and distributed-memory computers as well as on networks of workstations being used as parallel computers. The semantics of `eval` require that the child processes explicitly restore all state information after they are created. Linda allows a small number of simple data items to be passed in the argument list of the created processes; all other data must be passed through TS. This state information includes all shared-memory arrays, all COMMON blocks, and possibly, local static data.

Implementation

Analysis Of State Information. In evaluating the shared-memory version of Prism, two primary issues had to be addressed: array existence and array reference. While studying the dynamic array allocation in routines that call Prism, it was noted that many arrays were equivalenced to one another, or equivalenced to a local (scalar) scratch variable. The distinct existence of any particular array depended on which link was being considered and on user-specified options defining the type of results desired. Understanding all the permutations of links and options was crucial to developing a robust implementation.

The second issue arose in studying actual array references in Prism and in the routines called by Prism. It was noted that under certain combinations of options, many arrays were never referenced. This suggested that these arrays would not have to be passed through TS for certain types of jobs. This was the only effort made to reduce the amount of data passed through TS. Similar conclusions were reached while studying ChewER.

The other types of state information considered were COMMON blocks and local static data. It was determined that Prism needed six COMMON blocks initialized, while ChewER required 13. A thorough study of all the routines possibly called by Prism and ChewER showed several contained local static data, in the form of local variables appearing in SAVE statements. Many of these routines were low-level general-purpose Gaussian support routines. Each routine was considered individually, and it was determined that the effect of the variables reverting to their initial values was benign. Thus, there was no need to pass these variables through TS.

Code Structure. The above analysis led to an implementation that required only minor changes to the existing Gaussian code. A new version of the Prism setup routine PrsmSu was written. It receives over 100 arguments destined for Prism, and parcels them out as follows. Thirty integer scalars are stored into a local integer array (INTARY), which is then deposited into TS. The 6 required COMMON blocks are then deposited into TS. Depending on various problem-dependent options, as many as 45 arrays are then deposited.

Next, PrsmSu starts the parallel processes using eval. The entry point for each of these concurrent processes is a new routine named PrsmEv, described below; its arguments are ten logical scalars and two integer scalars. Following the initiation of the processes, PrsmSu calls Prism, thus participating in the parallel solution. When this invocation of Prism returns, PrsmSu gathers the results from its parallel processes, using the in operation. Each result array is then added into the corresponding array in PrsmSu. Finally, PrsmSu uses in to remove any data still in TS; this is merely a cleanup step to prepare TS for its next use.

The routine PrsmEv behaves like a MAIN routine in standard Fortran, except that it receives some arguments. It is responsible for getting all the state information from TS, allocating memory, calling Prism, and depositing the results into TS. First, it uses rd to get a copy of INTARY, the elements of which are stored into local scalar variables. Next, it uses rd to get copies of the COMMON blocks. At this point PrsmEv has all the control options required to calculate the number of required arrays (both input and output) and their sizes, so it can use malloc() to acquire the memory needed to hold them. It then uses rd to fill the input arrays, and initializes the output arrays to zero. This completes the necessary initialization, and Prism is called. When Prism returns, PrsmEv deposits the result arrays into TS, calls free() to release its memory, and exits.

The implementation for ChewER was constructed along similar lines, although the number and sizes of the various arguments differ significantly.

One final implementation detail should be mentioned. A single Gaussian run can execute dozens of links, possibly invoking several that have been parallelized.

Because each link is built as a stand-alone executable program, this implies (for a network version) that `ntsnet` would have to be executed several times. That is, `ntsnet` cannot initiate the Gaussian execution chain; instead, `ntsnet` must be initiated as part of the chain. This was accomplished by writing a C program called `lnklnd`, which is symbolically linked to the names of the parallelized Gaussian links. This program calls `system()` to create a shell in which `ntsnet` can run the actual parallelized link. The call to `system()` does not return until the shell exits, i.e., until `ntsnet` has terminated. At that time, `lnklnd` initiates the next link in the execution chain.

Parallel Results

Procedure. Timing runs were conducted using three variations of a test case from the Gaussian test suite. The molecule considered was triamino-trinitro-benzene (TATB), $C_6H_6N_6O_6$. With an appropriate choice of basis set, this job is not especially large by Gaussian standards, but it does represent a reasonable lower bound to problems that are large enough to warrant parallel processing. All tests were closed-shell calculations using direct methods.

Speedups and efficiencies were calculated as follows:

$$s = t_p / t_1$$
$$e = (s / p) * 100$$

where s is speedup, e is efficiency, t_p is elapsed time for the parallel calculation, t_1 is the elapsed time for the sequential calculation, and p is the number of processes. The single-process jobs were run using the standard (sequential) version of Gaussian, modified by the addition of the timing calls mentioned below.

Elapsed wallclock time (in seconds) was measured in two different places. First, library calls were added to `PrsmSu` and `CherSu` to record the elapsed wallclock time of the parallelized Prism and ChewER, including all process initiation and data communication costs. These times were used to calculate speedups and efficiencies for the parallelized pieces of code. Second, the Unix utility `date` was used before and after the Gaussian execution, to measure total elapsed wallclock time for the entire job. This includes both parallel and sequential time, including I/O and program (link) initiation times. Speedups and efficiencies for the complete start-to-finish jobs were then computed. Detailed tabular results may be found in the Appendix; below we present our results graphically.

Development and testing were initially performed on a Silicon Graphics Iris 4D/320GTX, a two-CPU shared-memory multiprocessor. The same program was then recompiled with Network Linda and timed on an ethernet-connected network of six IBM RS6000 Model 560 workstations. In this environment, all workstations were dedicated during the timing runs, although the network was not electrically isolated.

It should be noted that all testing was performed using the most recent development version of Gaussian, and should not be taken as indicative of the performance of Gaussian 92 as it is distributed.

Test Case 1. For the first test, a Hartree-Fock single-point energy calculation on TATB was performed using the basis set 6-31G** (300 basis functions). The SCF solution was produced after eight iterations, i.e., PrsmSu was called eight times. The amount of data transferred through TS varied from 2 MB to 4 MB for each iteration.

Results for Test Case 1

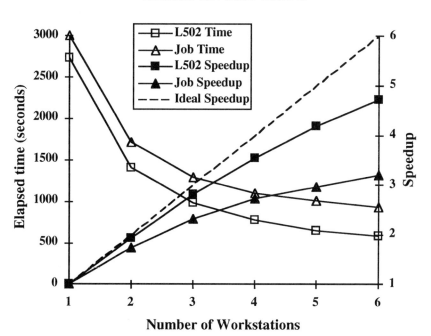

Number of Workstations

In this test case, link L502 is invoking Prism in order to form the Fock matrix during the SCF solution phase. Much of the drop in efficiency within Prism can be attributed to the redundant work in PickS4. The far greater decline in efficiency for the entire job is a combination of the redundant work, the remaining sequential calculations in link L502 and other links (especially link L401, which generates the initial guess of the density matrix), and poor network configuration. In particular, the residual sequential time increased from 260 seconds with one process to 358 seconds with six processes. This increase of almost 100 seconds is caused by slow ntsnet initialization on the workstation cluster being used. Subsequent testing has revealed a faulty automount mechanism on the particular workstation used as the master node. The system administrators for the network have not yet satisfactorily resolved this problem, so we have been unable to re-run our tests in an optimized network setting. However, we believe that most of the wasted time can be recovered through appropriate network configuration and tuning. In a feasibility study such as this one, we did not believe it necessary to pursue such tuning. Moreover, since this type of overhead is independent of problem size, it will be less significant for production-size calculations.

Test Case 2. The second test was a Hartree-Fock gradient calculation on TATB, also using the 6-31G** basis set. The SCF solution required 16 iterations to converge, due to the higher accuracy required by the subsequent gradient calculation. This test required between 2 MB and 4 MB of data to be passed through TS for each invocation of Prism.

Results for Test Case 2

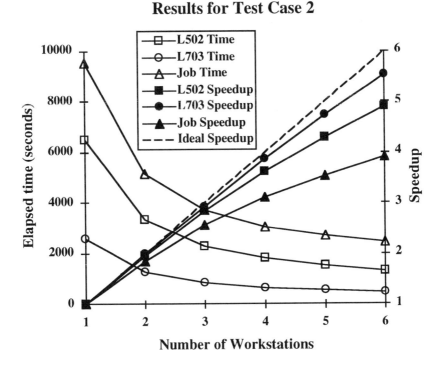

Here we see higher efficiencies for link L502 than in the previous test case. This is due to scaling effects; although the basis set contained the same number of functions, the higher accuracy necessary for convergence required more integrals to be computed. The effect of the redundant work in PickS4 is still present, but has less impact. For link L703, which here is invoking Prism in order to evaluate first derivatives, there is essentially no sequential calculation present, although there is still the redundant work in PickS4. The decline in Job efficiencies is due to the same causes as in the first test case, with the added (artificially high) cost of initializing ntsnet for both link L502 and link L703.

Test Case 3. The final test was a Hartree-Fock frequency calculation on TATB, this time using the 3-21G basis set (174 basis functions). The smaller basis set was chosen to allow the single-process run to complete in a reasonable amount of time; using the full 6-31G** basis, the single-process job would have required about two and one-half days of CPU time. While jobs of this magnitude are not unusual in Gaussian production environments, such resources were not available for this study. Unfortunately, the choice of the smaller basis has the undesirable effect of reducing the parallelizable computation. As with the previous test, link L502 took 16 iterations to converge to the SCF solution. The amount of data passed through TS varied widely, from around 800 KB (for the smallest L502 Prism iterations) to just over 60 MB (for the largest L1002 ChewER iterations). Due to the complexity of this test case, we shall present the speedup results separately from the elapsed time results.

Speedups for Test Case 3

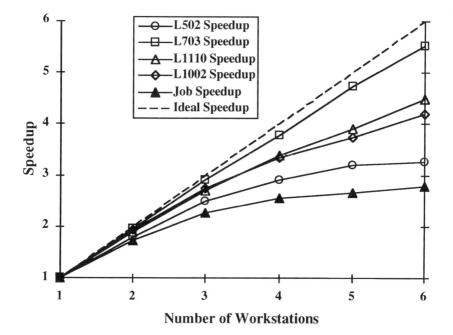

Number of Workstations

Comparing link L502 efficiencies with the previous test, we see the effect of the smaller basis. Link L703 (here evaluating second derivatives) still performs well, showing less sensitivity to basis set selection. This test case includes two additional parallelized links. Link L1110 performs well; it accomplishes more work in parallel than L502, computing integral derivatives and their contributions to 12 Fock matrices. As with all the other Prism links, L1110 suffers from the redundant work in PickS4. Link L1002 invokes ChewER to produce integral derivatives for the CPHF portion of the Fock matrix derivative calculation. The efficiencies for ChewER are good. They suffer from the redundant work of ChunkR, but should scale better than Prism because of the larger number of calculations being done in parallel. The Job efficiencies for this test are further reduced by the need for four invocations of ntsnet.

It is useful to look at the elapsed time results for this test case in a slightly different form. The following graph summarizes the elapsed time data.

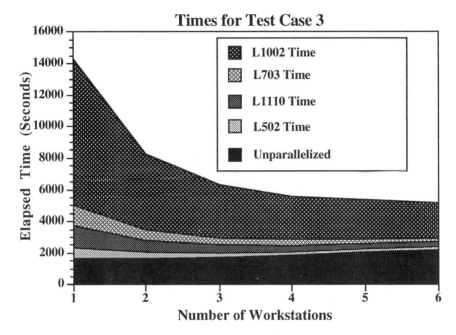

Times for Test Case 3

Legend:
- L1002 Time
- L703 Time
- L1110 Time
- L502 Time
- Unparallelized

Y-axis: Elapsed Time (Seconds)
X-axis: Number of Workstations

With one process, the elapsed time for the entire job was 14278 seconds; with six processes, the elapsed time was 5140 seconds, yielding a job efficiency of 46.3%. It is obvious that in the sequential calculation, the execution time is dominated by the CPHF (L1002) calculation. However, with six processes, the CPHF time is essentially the same as the unparallelized time. This suggests that future efforts need to focus on additional parallelization as well as on improving the existing parallel performance.

Future Work

In addition to constructing an experimental network-parallel Gaussian and producing the timing results presented above, we also conducted a number of profiling studies to identify areas for future parallelization efforts. These studies identified a number of matrix arithmetic operations which need to be parallelized. For example, there is a matrix diagonalization following each Prism invocation during the SCF iteration. Furthermore, there are numerous matrix multiplication routines which are heavily used. Some of these are invoked with a large number of (relatively) small but independent matrices; others are invoked with a small number of very large matrices. Appropriate strategies will have to be employed depending on the granularity of the particular operation under consideration.

Profiling was also performed on two of the more accurate (and expensive) post-SCF methods available in Gaussian. The results for direct and semi-direct MP2 and semi-direct QCISD(T) were consistent with those above, indicating a need to parallelize basic matrix operations such as multiplication. However, the situation with these algorithms is complicated by the fact that they are disk-based algorithms, due to the enormous size of the matrices involved. However, since the aggregate memory available on a network is often far larger than that on a single machine, it may be possible to reduce the I/O in a parallel version. Even so, it will still be necessary to exercise care in implementing these matrix operations to avoid any I/O bottleneck.

Since the completion of this work, we have had an opportunity to experiment with some extensions that fall beyond the scope of our original intentions. Specifically, we have applied the parallel model used for Prism and ChewER to Gaussian's density functional theory (DFT) model. Also, we have experimented with a "server" version of the parallel model in an attempt to reduce startup costs. Preliminary results from both of these efforts look promising, and we hope to report more fully on them in the future.

Conclusions

This work has clearly demonstrated the feasibility of developing a version of Gaussian capable of significant parallel speedups on a network of high-performance workstations. By using the Linda parallel programming environment, development and debugging efforts were minimized, while the underlying portability and performance of Gaussian were maintained. Potential obstacles to increased parallel performance were identified. Some minor tuning and implementation issues were also identified; these need to be resolved in order to fully exploit the currently available parallelism.

References

1. Frisch, M. J.; Trucks, G. W.; Head-Gordon, M.; Gill, P. M. W.; Foresman, J. B.; Wong, M. W.; Johnson, B. G.; Schlegel, H. B.; Robb, M. A.; Replogle, E. S.; Gomperts, R.; Andres, J. L.; Raghavachari, K.; Binkley, J. S.; Gonzalez, C.; Martin, R. L.; Fox, D. J.; Defrees, D. J.; Baker, J.; Stewart, J. J. P.; Pople, J. A.; Gaussian, Inc.: Pittsburgh, 1992
2. Schlegel, H. B.; Frisch, M. J. In *Theoretical and Computational Models for Organic Chemistry*; S. J. Formosinho, Ed.; Kluwer Academic Pubs.: The Netherlands, 1991; pp 5-33.
3. Gill, P. M. W.; Head-Gordon, M.; Pople, J. A. *J. Phys. Chem.* **1990**, *94*, 5564-5572.
4. Gill, P. M. W.; Johnson, B. G.; Pople, J. A. *Int. J. Quant. Chem.* **1991**, *40*, 745-752.
5. Gill, P. M. W.; Pople, J. A. *Int. J. of Quantum Chem.* **1991**, *40*, 753-772.
6. Frisch, M. J.; Head-Gordon, M.; Pople, J. A. *J. Chem. Phys.* **1990**, *141*, 189-196.
7. *Modern Techniques in Computational Chemistry*; Clementi, E., Ed.; Escom Science Publishers: 1990.
8. Colvin, M.; Janssen, C. *MPCRL Research Bulletin* **1993**, *3*, 6-9.
9. Dupuis, M.; Watts, J. D. *Theoret. Chim. Acta* **1987**, *71*, 91.
10. Guest, M. F.; Harrison, R. J.; vanLenthe, J. H.; vanCorler, L. C. H. *Theoret. Chim. Acta* **1987**, *71*, 117.
11. Luthi, H. P.; Mertz, J. E.; Feyereisen, M. W.; Almlof, J. E. *J. Comp. Chem.* **1992**, *13*, 160-164.
12. Feyereisen, M. W.; Kendall, R. A.; Nichols, J.; Dame, D.; Golab, J. T. *J. Comp. Chem.* **1993**, *14*, 818-830.
13. Frisch, M. J.; Gomperts, R. *paper in preparation*

14. Scientific Computing Associates Inc. *Fortran-Linda Reference Manual*; New Haven, CT: 1993.
15. Arango, M.; Berndt, D.; Carriero, N.; Gelernter, D.; Gilmore, D. *Supercomputing Review* **1990**, *10*, 42-46.
16. Scientific Computing Associates Inc. *C-Linda User's Guide & Reference Manual*; New Haven, CT: 1993.
17. Carriero, N.; Gelernter, D. *How to Write Parallel Programs: A First Course*; MIT Press: Cambridge, 1990.
18. Bjornson, R. Ph.D. Dissertation Thesis, Department of Computer Science, Yale University, 1993.
19. Bjornson, R.; Carriero, N.; Gelernter, D.; Kaminsky, D.; Mattson, T.; Sherman, A. "Experience with Linda," Department of Computer Science, Yale University, 1991.
20. Carriero, N. "Implementation of Tuple Space Machines," Department of Computer Science, Yale University, 1987.

Appendix

In the tables below, **p** is the number of processes, **t** is the measured elapsed wallclock time in seconds, **s** is the computed speedup, and **e** is the computed efficiency.

Results for Test Case 1

	Prism (L502)			Job		
p	**t**	**s**	**e**	**t**	**s**	**e**
1	2739			2999		
2	1409	1.94	97.2	1719	1.74	87.2
3	977	2.80	93.4	1294	2.32	77.3
4	771	3.55	88.8	1104	2.72	67.9
5	653	4.19	83.9	1009	2.97	59.4
6	580	4.72	78.7	938	3.20	53.3

Results for Test Case 2

	Prism (L502)			Prism (L703)			Job		
p	**t**	**s**	**e**	**t**	**s**	**e**	**t**	**s**	**e**
1	6507			2565			9541		
2	3325	1.96	97.8	1292	1.99	99.3	5162	1.85	92.4
3	2288	2.84	94.8	874	2.93	97.8	3727	2.56	85.3
4	1801	3.61	90.3	663	3.87	96.7	3060	3.12	77.9
5	1516	4.29	85.8	544	4.72	94.3	2699	3.54	70.7
6	1327	4.90	81.7	465	5.52	91.9	2439	3.91	65.2

Results for Test Case 3

	Prism (L502)			Prism (L703)			Prism (L1110)		
p	t	s	e	t	s	e	t	s	e
1	713			1326			1378		
2	399	1.79	89.3	680	1.95	97.5	726	1.90	94.9
3	286	2.49	83.1	455	2.91	97.1	510	2.70	90.1
4	245	2.91	72.8	350	3.79	94.7	408	3.38	84.4
5	222	3.21	64.2	280	4.74	94.7	352	3.91	78.3
6	219	3.26	54.3	240	5.53	92.1	307	4.49	74.8

	ChewER (L1002)			Job		
p	t	s	e	t	s	e
1	9243			14278		
2	4783	1.93	96.6	8234	1.73	86.7
3	3364	2.75	91.6	6323	2.26	75.3
4	2755	3.35	83.9	5587	2.56	63.9
5	2480	3.73	74.5	5357	2.67	53.3
6	2204	4.19	69.9	5140	2.78	46.3

RECEIVED November 15, 1994

Chapter 6

The Parallelization of a General Ab Initio Multireference Configuration Interaction Program

The COLUMBUS Program System

Hans Lischka[1], Holger Dachsel[1], Ron Shepard[2], and Robert J. Harrison[3]

[1]Institut für Theoretische Chemie und Strahlenchemie, Universität Wien,
A-1090 Vienna, Austria
[2]Argonne National Laboratory, Argonne, IL 60439
[3]Pacific Northwest Laboratory, Richland, WA 99352

A massively parallel version of the diagonalization section of the the COLUMBUS MRSDCI program system is reported. Coarse grain parallelization is performed at the topmost level of the program by means of the segmentation of the trial and resulting update vectors of the iterative Davidson scheme. Message passing based on the portable toolkit TCGMSG and the global array tools are used for communication between processors. Test calculations with CI dimensions of more than 2.5 million were carried out on the Intel Touchstone Delta with a parallel efficiency of more than 90% on 320 processors. An outline of the parallelization of the entire program system is also given.

Parallel computing is one of the great challenges in the computationally oriented sciences. It is of particular importance in Quantum Chemistry since practically all computational methods are extremely time consuming if realistic simulations of molecules are attempted. Starting with the pioneering work by Clementi and coworkers on "loosely coupled array of processors (LCAP)" (1) several investigations on the parallelization of SCF programs have been reported (2-11). In addition, efforts to parallelize electron correlation methods like Møller-Plesset Second-Order Perturbation Theory (12), Coupled-Cluster theory (13,14) and full CI (15,16) have been undertaken as well. For reviews on the use massively parallel computers in Quantum Chemistry see e.g. (17,18).

Because of its simplicity, the direct SCF approach is most promising with respect to parallelization. The accuracy of the SCF method is sufficient for many standard situations in chemistry. However, if one wants to achieve higher precision and/or more general applicability electron correlation methods have to be used. Here, the situation is more complex since, depending on the case, at least a partial transformation of the two-electron integrals from the AO into the MO basis has to be carried out. One of the most complicated, but also one of the most general methods is the multi-reference single- and double-excitation configuration interaction (MRSDCI) approach. It is our aim to

0097–6156/95/0592–0075$12.00/0

develop a portable, massively parallel MRSDCI program on the basis of the COLUMBUS program system (19,20). In our previous investigation (21) we had shown that via coarse grain parallelization and message-passing based on the portable program package TCGMSG (22) developed by one of us (RJH) the diagonalization step can be very well parallelized. This program version was working on a variety of parallel computers, both shared memory (Alliant FX/2800, Cray Y-MP, Convex C2) and distributed memory (iPSC/860). Calculations with up to 8 processors were performed. The most severe bottlenecks of that version were located in the managment of the data files (mainly the two-electron integrals and the CI- and other expansion vectors) because of the restrictions inherent in the message-passing model. In order to improve this situation we have made use of the recently developed global-array tools (23) (see below). With these tools and introducing other enhancements (like virtual disks and an improved dynamic load balancing scheme) we are now in the position to use efficiently more than 300 processors on the Intel Touchstone Delta. It is the purpose of this communication to report in more detail on our achievments and to give an outline of our plans concerning the parallelization of the entire program system.

Review of our Previous Work

The COLUMBUS program system (19,20) is a collection of Fortran programs for performing general *ab initio* MRSDCI calculations and is based on the Graphical Unitary Group Approach (24,25). For accurate, large scale MRSDCI calculations the computationally most demanding section is the diagonalization of the matrix representation of the hamiltonian operator in the basis of the configuration state functions (CSFs). Expansion lengths of 1 - 10 million are now becoming routine with the COLUMBUS program system. The iterative Davidson diagonalization method (26) is used to determine the appropriate eigenvectors and eigenvalues. In this scheme, the most important step by far is the computation of a matrix-vector product $\mathbf{w}_i = \mathbf{H}\mathbf{v}_i$ ($\mathbf{w}_i$ is called the resulting product vector) of the hamiltonian matrix $\mathbf{H}$ and trial vectors $\mathbf{v}_i$. A "direct CI" procedure (27) is used to compute this matrix-vector product. It is driven by the four indices of the two-electron integrals. In order to compute the subspace representation of $\mathbf{H}$ with respect to the trial vectors and the overlap matrix for the trial vectors scalar products $\mathbf{v}_i^+\mathbf{w}_j$ and $\mathbf{v}_i^+\mathbf{v}_j$ have to be calculated as well.

As has been discussed in detail in (21) we decided for a coarse grain parallelization of $\mathbf{H}\mathbf{v}$ in which the outermost loops over segment pairs of the vectors $\mathbf{v}$ and $\mathbf{w}$ were used for parallelization. This choice had several advantages:

- it provided the most coarse-grain decomposition possible,
- the complexity of the code below these loops did not affect the parallelization,
- each process only needed to hold four vector segments,
- the number of tasks is actually proportional to the square of the number of segments. Thus it is possible to generate a sufficiently large number of tasks to make load balancing effective.

Dynamic load balancing was implemented via a shared counter. Each index in this counter corresponded to a task defined by the work to compute the contribution of one segment pair to the matrix-vector product $\mathbf{w} = \mathbf{H}\mathbf{v}$. Each process had a local copy of $\mathbf{w}$ on which the partial contributions to it were updated. Each process also had to read all integrals (and other quantities, like indexing arrays, which were not important in terms of I/O). After completion of the loops over segment pairs the total vector **w** was obtained by a global sum operation from all partial contributions which had been accumulated by each process. In the original, sequential program there was also a formula tape which contained the coupling coefficients which determined the contribution of the one- and two-electron integrals to each matrix element of **H**. The formula tape entries were determined in terms of internal MO indices only. This formula tape was replaced by recalculating the required coupling coefficients on the fly. The actual updating scheme was carried out in terms of dense-matrix kernels (BLAS routines (28)) of the dimension of the MO basis. All files were located on magnetic disk and were accessed in the same way as in the sequential program. The subspace manipulations were not parallelized at all.

Because of the top level coarse-grain parallelization only few and rather simple modifications had to be implemented into the original sequential code. In particular, the above-mentioned updating scheme in terms of dense-matrix kernels was completely unchanged.

Outline of the New Features

The main purpose of our first implementations was to investigate the overall efficiency of our segmentation scheme. From the analysis of the timings reported in (21) one can see that load balancing worked very well for the number of processors used and that substantial speedups could be achieved. However, it was also clear from the results obtained especially from the iPSC/860 that storing data on disk and having each processor access them directly from there created a severe bottleneck. In order to judge the amount of data transferred it is important to note that the 4-,3- and 1-external integrals were only read $N_{seg}/2$ times (N_{seg} being the number of segments) and the remaining integrals (2- and 0-external) $N_{seg}(N_{seg}-1)/2$ times in each iteration in the Davidson procedure. Thus, for the test cases and the number of segments (typically between 20 to 30) chosen in (21) the amount of data transfer was dominated by the 2-external integrals even though they only constitute a small fraction of the 4- and 3-external ones. Also, reading the trial vector segments and writing the resulting product vector segments for each segment pair created substantial I/O overhead.

In order to overcome the just mentioned difficulties we proceeded in two steps. In the first step we tried to reduce the amount of data transfer to a minimum while still using the conventional message-passing tools. In the second step we extended message-passing by the global-arrays tools in order to allow more flexibility in accessing data distributed over the memory of the individual processors.

For the purpose of comparison with our previous timings we use the same CH_3 test cases as before (21): C_{2v}-pVDZ and C_1-pVTZ. The CI dimensions were 70 254 and 2 528 400, respectively.

Virtual Disk and Data Compression. First of all, we introduced the concept of a virtual disk residing in central memory. Files could be optionally written to this virtual disk instead of writing them to a magnetic disk. Thus, slow disk I/O was replaced by fast internally copying data in central memory. Since only space for four segments (and some other data buffers, etc.) have to be kept in core at the same time the requirements of our program concerning central memory are rather modest. Therefore, depending on the actual central memory available on a given computer, we can set aside an additional amount of memory for the purpose of a virtual disk. Overflow of this storage area to disk is possible so that in cases of insufficient central memory the calculation does not break down but continues with ordinary disk I/O.

In order to make use of the virtual disk in an economic way a data compression scheme for the trial and the resulting product vectors was developed (29). Each of these vectors was truncated to a fixed number of decimal places which is chosen such that a given accuracy in the Davidson diagonalization scheme is obtained. A special floating point representation with a 7 bits exponent, one bit for the sign and with a variable length mantissa is used for that purpose. With an energy threshold of 10^{-6} hartree an overall reduction of the two vector files by factors between 4 to 6 is achieved. Moreover, the subspace dimension in the Davidson procedure is set to four in order to reduce also in this way the amount of data to be stored. A new trial and resulting product vector is constructed from the individual subspace components each time the limit of the subspace expansion is reached.

Test calculations were performed on an iPSC/860 with 16 MB available on each node. The C_{2v}-pVDZ test case is small enough (integral files 1.2 MB total, compressed trial vectors and resulting product vectors 1 MB, compression factor 4.5) so that all files can be kept as separate copies on the local virtual disk of each processor. Since the integral files do not change during the iterations they are copied to the individual nodes at the beginning of the calculation. In each Davidson iteration, the trial vector is broadcast to all nodes and after completion of the hamiltonian matrix trial vector product the partial results are summed up via a global sum operation. The subspace manipulations were not parallelized at this stage of program development. Using the above-mentioned test example no disk I/O was performed at all during the entire calculation. This replicate data approach gave us well defined conditions in order to study the performance of the dynamic load balancing scheme in detail. Of course, it is not well suited to allow larger calculations because of its extensive memory requirements.

In Table I timings for the iPSC/860 based on the just described program version are given. Compared to the previous results reported in Tab. 5 of Ref. 21 significant improvements in the speedups for the Hv step are observed. In particular, speedup factors of 6.8 for the 8 processor case and 11.9 for the 16 processor case are found. The difference between the observed and theoretical values are due to deficiencies in the load balancing scheme in that particular program version. In the timings for the global sum operation cpu times for data compression are included as well. As noted above, the subspace manipulations were not parallelized.

Table I. Timings for the CH₃ C_{2v}-pVDZ test case determined on the iPSC/860[a]

no. procs.	1	2	4	8	16
1. broadcast	0.0	1.6	1.7	1.7	1.7
2. Hv	303.6	159.0	83.5	44.5	25.5
3. global sum	5.8	6.3	7.2	8.7	10.2
4. subspace	4.7	4.7	4.7	4.7	4.7
5. complete iteration	314.2	173.0	97.1	60.0	42.4

[a] Timings are given in secs. wall clock time.

Global Arrays and Improved Dynamic Load Balancing. Despite of the advantages of the replicate data approach described above there are some major drawbacks to it. One has already been mentioned and comes from excessive memory requirements if one wants to keep local copies of the integral, v and w files on each node. Another one arises because the entire trial vector has to be broadcast to all processors even though only a fraction of it will be used. Since we use dynamic load balancing, we do not have a deterministic sequence of tasks on each processor. Therefore, the whole vector has to be communicated at the beginning of each iteration. At the end of each iteration the partial updates to the resulting product vector **w** computed by each process have to be summed up resulting in a synchronization barrier. There is ample opportunity in the program to interleave asynchronous reading data from and writing data to individual nodes without interfering significantly with the operations taking place on other nodes. E.g., it is not necessary to read the entire trial vector **v** at the beginning of the iteration. It would be sufficient to access just those segments at the time they are actually needed. Similar arguments apply to the updating procedure of the vector **w**.

In order to achieve this increased flexibility we use the global array toolkit developed by one of us (23). These tools support one-sided access to data structures (here limited to one- and two-dimensional arrays) in the spirit of shared memory. With some effort this can be done portably resulting in a much easier programming environment, speeding up code development and maintainability. Significant performance enhancements are observed by the aforementioned utilization of asynchrony of the execution of processes.

Using the global arrays we are in the position to distribute all major files over the memory of the individual nodes. It is stressed, that in contrast to the previous program version now no multiple copies of files are required. However, we can still keep any files additionally, if we wish, as multiple copies as before in the aforementioned local virtual disks on each processor. This is especially advisable in cases, like the two-external integrals, where the file size is moderate but the file is read frequently. No access to disk files is made during the calculation.

With the global arrays it is also straightforward to parallelize the subspace manipulations as well. Since the segment distribution over processors is the same for **v** and **w** the computation of the scalar products can be arranged such that each processor only accesses local vector segments. Only these partial contributions for each segment have to be summed globally. The necessary communication between processors for that purpose is very small. Overall, the subspace step is very well parallelized and does not contribute significantly to the total timings.

When going to larger numbers of processors the flexibility of our load balancing scheme had to be increased as well. Because of Amdahl's law (30) it is of utmost importance to avoid idle times of processors. Such a situation arises because some of the processors finish the **Hv** step earlier than others since no more work is available. In order to avoid this idling the granularity of the tasks towards the end of the computation of **Hv** has to be sufficiently small. In the original program each task consisted of the work for one pair of segments of **v**. First of all, it was straightforward to further subdivide this task with respect to the number of internal orbital indices of the two-electron integrals. A task list was generated and ordered according to decreasing timings (at the moment determined from the timings of the first iteration). The load balancing via the shared counter was now based on that ordered task list which usually gave a nearly optimal load balance. In order to quantify the amount of idle times we define an efficiency of parallelization e for the **Hv** step as

$$e = \frac{t_{total}}{nproc \cdot t_{max}}. \tag{1}$$

t_{total} is the total time needed by all processors to compute **Hv**, t_{max} is the longest time on one processor and $nproc$ is the number of processors.

Benchmark Calculations

Benchmark calculations have been performed on the Intel Touchstone Delta at Caltech using the C_1-pVTZ test case. In order to give an overview of the space requirements for the global arrays individual file sizes are given as follows:

integrals
       4-external 38.0 MB, 3-external 16.3 MB, 2-external 1.4 MB, 1-external 0.13
       MB and 0-external 0.033 MB
v and **w** vectors (4 vectors each)
       162.0 MB; diagonal elements of **H**- 20.3 MB plus several indexing arrays 3.3
       MB

     The 2-, 1- and 0-external integrals and an indexing vector referring to internal walks (0.9 MB) were stored as separate copies on each node on local virtual disks (2.5 MB total). The remaining data (239.9 MB) were kept via global arrays as a single copy. Due to the very limited central memory available on each node of the Delta (12 MB including operating system) only about 1.5 MB were available for global arrays on each node. Thus, we need at least the memory of about 160 processors to accomodate the space for the global arrays. The trial and update vectors were split into 115 segments.
     In Fig.1 the speedup curve for calculations up to 512 nodes is shown and compared to the ideal behavior. The observed speedup curve follows the theoretical one very closely to about 320 processors. Then the speedup starts to deteriorate. From 448 to 512 processors even a decrease is to be observed. The reason for that behavior

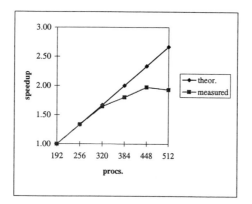

Figure 1. Speedup curves obtained on the Intel Touchstone Delta for the CH_3 C_1-pVTZ test case.

comes from the fact that some of the 3- and 4-external integral cases start to dominate the calculation. As already mentioned above, some processors are still busy while others are already idling because of lack of work. This behavior is cleary demonstrated in Fig. 2 where the efficiency of parallelization as defined in equation 1 is shown in dependence of the number of processors. The efficiency of parallelization remains well above 90% up to 320 processors and then drops significantly for the just mentioned reasons. It would not be too difficult to split the 3- and 4-external cases further. However, since we plan to formulate these cases in a totally different way in the next future (see below) we renounced in spending some effort into this aspect of program optimization now. We also want to stress that even though we did not perform the C_1-pVTZ test case with less than 192 processors, from our experience with other test runs (on workstation clusters and on the IBM SP1) we are positive that the respective speedups would also follow closely the theoretical curve to lower processor numbers if only more central memory would be available on each node.

Conclusions and Outlook

Using the global-array toolkit a very satisfactory parallelization of the CI part of the COLUMBUS program system has been achieved. We are in the position to run that most complicated and in many cases also by far dominating part very efficiently on more than 300 nodes on the Intel Touchstone Delta. Implementations on the IBM SP1 and the KSR2 are planned for the next future. As the program is now, the computer time increases significantly with the number of segments (21) because some overhead is introduced in the formula generation for each segment pair. In order to reduce this overhead we have developed a scheme which uses specially adjusted Distinct Row Tables (for the definition of the DRT and further information on GUGA see (24,25)) in

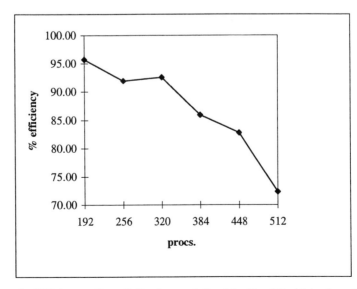

Figure 2. Efficiency of parallelization as defined by Eq. (1) obtained on the Intel Touchstone Delta for the CH_3 C_1-pVTZ test case.

order to construct just those coupling elements which are needed for that particular segment pair. First results are very encouraging.

The major next change in the program will affect the treatment of the 3- and 4-external integrals and is crucial for the design of the parallel structure of the whole program system. As it is now, the COLUMBUS program is structured in the conventional way, i.e. the two-electron integrals are calculated in the AO basis, transformed to the MO basis and then sorted into the different cases according to the number of external indices (0- to 4-external). These steps require a lot of complicated sorting and I/O steps with large amounts of data. This procedure is certainly not appropriate for parallelization. In case of a MRSDCI wave function AO driven formalisms for the 3- and 4-external integral cases have been developed (31,32). Therefore, a restricted integral transformation for the remaining 2-, 1- and 0-external integrals is required only. A first sequential code along these lines has been introduced into the COLUMBUS program as well (33). It will be the basis for a "double direct" MRSDCI program where all major I/O bottlenecks will be removed and which should be particularly well suited for parallelization.

Acknowledgments

This work was performed under the auspices of the Austrian "Fonds zur Förderung der wissenschaftlichen Forschung", project nr. P9032 and the High Performance Computing and Communication Program of the Office of Scientific Computing and the Office of Basic Energy Sciences, Division of Chemical Sciences, U.S. Department of Energy, under the contract number W-31-109-Eng-38 with the Argonne National Laboratory and under contract DE-AC-76RLO 1830 with Battelle Memorial Institute which operates the Pacific Northwest Laboratory. The calculations on the iPSC/860 and the Intel Touchstone Delta were performed at the CCSF at Caltech, those on the IBM SP1 at the ACRF of the Argonne National Laboratory. We are grateful for the competent support of our work by these computer centers.

References

1. E. Clementi, *Modern Techniques in Computational Chemistry*, E. Clementi, Ed., Escom Science Publishers, 1990, chap. 1; D. Folsom *Modern Techniques in Computational Chemistry*, E. Clementi, Ed., Escom Science Publishers, 1990, chap. 27.
2. Dupuis, M., Watts, J. D. *Theor. Chim. Acta* **1987**, *71*, 91.
3. Harrison, R. J., Kendall, R. A. *Theor. Chim. Acta* **1991**, *79*, 337.
4. Lüthi, H. P., Mertz, J. E., Feyereisen, M. W., Almlöf, J. E. *J. Comp. Chem.* **1992**, *13* 160.
5. Kindermann, S., Michel, E., Otto, P. *J. Comp. Chem.* **1992**, *13*, 414.
6. Feyereisen, M. W., Kendall, R. A., Nichols, J., Dame, D., Golab, J. T. *J. Comp. Chem.* **1993**, *14*, 818.
7. Brode, S., Horn, H., Ehrig, M., Moldrup, D., Rice, J. E., Ahlrichs, R. *J. Comp. Chem.* **1993**, *14*, 1142.
8. Schmidt, M. W., Baldridge, K. K., Boatz, J. A., Elbert, S. T., Gordon, M. S, Jensen, J. H., Koseki, S., Matsunaga, N., Nguyen, K. A., Su, S., Windus, T. L., Dupuis, M., Montgomery, J. A., Jr. *J. Comp. Chem.* **1993**, *14*, 1347.
9. Colvin, M. E., Janssen, C. L., Whiteside, R. A., Tong, C. H. *Theor. Chim. Acta* **1993**, *84*, 301.

10. M. Petterson, L. G., Faxen, T. *Theor. Chim. Acta* **1993**, *85*, 345.
11. Burkhardt, A., Wedig, U., Schnering, H. G. v. *Theor. Chim. Acta* **1993**, *86*, 497.
12. Watts, J. D., Dupuis, M. *J. Comp. Chemistry* **1988**, *9*, 158.
13. Rendell, A. P., Lee, T. J., Lindh, R. *Chem. Phys. Lett.* **1992**, *194*, 84.
14. Rendell, A. P., Guest, M. F., Kendall, R. A. *J. Comp. Chem.* **1993**, *14*, 1429.
15. Harrison, R. J., Stahlberg, E. A. *J. Parallel and Distributed Computing* in press.
16. Bendazzoli, G. L., Evangelisti, S. *J. Chem. Phys.* **1993**, *98*, 3141.
17. Colvin, M. E., Whiteside, R. A., Schaefer III, H. F. *Methods in Computational Chemistry*; Wilson, S., Ed., Plenum: N.Y., 1989, Vol. 3, pp. 167.
18. Kendall, R. A. *Int. J. Quantum Chem.* **1993**, *S27*, 769.
19. Lischka, H., Shepard, R., Brown, F., Shavitt, I. *Int. J. Quantum Chem.* **1981**, *S15*, 91.
20. Shepard, R., Shavitt, I., Pitzer, R. M., Comeau, D. C., Pepper, M., Lischka, H., Szalay, P. G., Ahlrichs, R., Brown, F. B., Zhao, J. G. *Int. J. Quantum Chem.* **1988**, *S22*, 149.
21. Schüler, M., Kovar, T., Lischka, H., Shepard, R., Harrison, R. J. *Theor. Chim. Acta* **1993**, *84*, 489.
22. Harrison, R. J. *Intern. J. Quantum Chem.* **1991**, *40*, 847.
23. Harrison, R. J. *Theor. Chim. Acta* **1993**, *84*, 363 and unpublished further work.
24. Paldus, J. *The Unitary Group for Evaluation of Electronic Energy Matrix Elements*; Hinze, J., Ed., Springer-Verlag: Berlin 1981, pp.1.
25. a) Shavitt, I. *The Unitary Group for Evaluation of Electronic Energy Matrix Elements*, Hinze, J., Ed., Springer-Verlag: Berlin 1981, pp.51.
    b) Shavitt, I. *Mathematical Frontiers in Computational and Chemical Physics*; Truhlar, D. G., Ed., Springer-Verlag: Berlin 1988, pp. 300.
26. Davidson, E. R. *J Comp Phys* **1975**, *17*, 84.
27. a) Roos, B. O., *Chem. Phys. Lett.* **1972**, *15*, 153.
    b) Roos, B. O., Siegbahn, P. E. M. *Methods of Electronic Structure Theory*; Schaefer III, H. F., Ed., Plenum: NY 1977, pp. 277.
28. a) Dongarra, J. J., DuCroz, J., Hammerling, S., Hanson, R. *ACM Trans. on Math. Soft.* **1988**, *14*, 1.
    b) Dongarra, J. J., DuCroz, J., Duff, I., Hammerling, S. *ACM Trans. on Math. Soft.* **1990**, *16*, 1.
29. Dachsel, H., Lischka, H. unpublished results.
30. Amdahl, G. M. *Proc. AFIPS Spring Joint Computer Conf.* **1967**, *30*, 40.
31. Werner, H. J., Reinsch, E.-A. *J.Chem.Phys.* **1982**, *76*, 3144.
32. Ahlrichs, R. *Proceedings of the 5th Seminar on Computational Methods in Quantum Chemistry*, Duijnen, T.H.v., Nieuwport, W.C., Eds., MPI Garching, Germany 1982.
33. Kovar, T., Lischka, H. unpublished results.

RECEIVED January 19, 1995

Chapter 7

Parallel Calculation of Electron-Transfer and Resonance Matrix Elements of Hartree–Fock and Generalized Valence Bond Wave Functions

Erik P. Bierwagen, Terry R. Coley, and William A. Goddard, III

Materials and Molecular Simulation Center, Beckman Institute, California Institute of Technology, 139-74, Pasadena, CA 91125

We review the theory for the computation of the Hamiltonian matrix element between two distinct electronic wave functions Ψ_A and Ψ_B sharing the same nuclear configuration but differing electronic density distributions. For example, Ψ_A and Ψ_B might describe two endpoints in an electron transfer reaction or two configurations in a resonance description of a molecule. In such cases the calculation of the rate of electron transfer or resonance energy requires evaluation of $\langle \Psi_A | \hat{H} | \Psi_B \rangle = H_{AB}$ matrix elements. Because the orbitals of Ψ_A and Ψ_B have complicated (non-orthogonal) relationships, the calculation of H_{AB} had been computationally intensive. In this paper we consider Ψ_A, Ψ_B having the form of closed or open-shell Hartree-Fock or Generalized Valence Bond wave functions and show the parallel structure of the theory. Using this parallel structure we present an efficient computational implementation for shared memory multiprocessors.

The starting point of most *ab initio* quantum chemistry is an antisymmetrized product of molecular orbitals $\Psi_A = |\phi_a \phi_b \phi_c \cdots|$. To compute properties such as energy, $E_A = \langle \Psi_A | \hat{H} | \Psi_A \rangle$, the molecular orbitals of Ψ_A are constructed to be mutually orthogonal. However, many problems are conveniently described in terms of two different ground state wave functions. For example to describe the charge transfer between Ψ_A and Ψ_B:

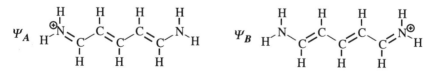

we need to compute cross-matrix elements, $\langle \Psi_A | \hat{H} | \Psi_B \rangle = H_{AB}$, where each molecular orbital of Ψ_A overlaps some or all orbitals of Ψ_B. In this case the electron transfer rate is proportional to $|T_{AB}|^2$, where:

$$T_{AB} = \frac{H_{AB} - S_{AB}H_{AA}}{1 - S_{AB}^2} \tag{1}$$

and $S_{AB} = \langle \Psi^A | \Psi^B \rangle$ is the overlap matrix.

Another example is the computation of chemical resonance energies. In this case Ψ_A and Ψ_B describe two different valence states (*e.g.*, the two valence states of benzene). Representing the resonating wave function as $\Psi = \Psi_A + \Psi_B$, we can calculate its energy,

$$E_{AB} = \frac{\langle \Psi^A + \Psi^B | \hat{H} | \Psi^A + \Psi^B \rangle}{\langle \Psi^A + \Psi^B | \Psi^A + \Psi^B \rangle} = \frac{H_{AA} + H_{BB} + 2H_{AB}}{2 + 2S_{AB}}, \tag{2}$$

only if we have the means to calculate H_{AB} and S_{AB}. Because of its intimate relationship to resonance energies, we will refer to H_{AB} as a resonance matrix element.

The computation of resonance matrix elements can also be used to evaluate configuration interaction (CI) wave functions in cases where the configurations are non-orthogonal. Such non-orthogonal CIs have been successfully carried out (*1*); however, the computational complexity has limited the applications. If they can be made practical, non-orthogonal CI approaches have two distinct advantages over orthogonal CIs:

1) the component states Ψ_A and Ψ_B can be chosen to be chemically meaningful descriptions of the system
2) this "better" choice of basis states reduces the number of states needed to accurately describe the system

Electronic reorganization problems such as electron transfer and interpretation of photoelectric spectra lead naturally to a few-state description in terms of non-orthogonal basis states.

A straightforward calculation of H_{AB} for non-orthogonal wave functions involves non-orthogonal matrix elements involving all orbitals of Ψ_A overlapping all orbitals of Ψ_B leading to an N! dependency, where N is the number of occupied spatial orbitals in each wave function. This contrasts with the case of orthogonal spatial orbitals where there are only of order N^2 operations. To simplify this problem Voter and Goddard (*2*) showed that a pair of unitary transformations exists, which when applied to the molecular orbitals of Ψ_A and Ψ_B, respectively, a) leave the total energy, E_{AB}, unchanged and b) reduce the computational effort to order N^2 by transforming Ψ_A and Ψ_B such that each orbital of Ψ_A overlaps exactly one orbital of Ψ_B. By reducing the computational effort this *biorthogonalization*, makes the resonance calculation tractable.

Despite the computational savings obtained with clever transformations such as biorthogonalization, many systems of interest, especially in electron transfer studies, remain too large for practical H_{AB} calculations with existing computer codes. Current programs, which have served well for smaller cases, do not exploit the underlying parallelism in the theory and therefore cannot take advantage of multi-processor computers without significant restructuring (*3*).

Goals Through the use of modern programming languages and software design we have produced a program for computing resonance matrix elements for systems of potentially unlimited size. The program meets the following design goals:
  • efficient performance on shared-memory multi-processors
  • user level control over the program's internal data structures and algorithms
The Method section exposes the parallelism inherent in the theory of resonance matrix element calculations. The Algorithm section introduces algorithms for shared-memory multi-processors and shows how the first goal was achieved. Program Architecture discusses our second goal in more general terms. Finally, in Results and Discussion we present timings and resonance energies for two systems of chemical interest.

Method

The computational theory of resonance matrix elements was developed by Voter and Goddard to examine the resonance energy between valence bond (and generalized valence bond (GVB)) (*4-7*) wave functions and is described elsewhere (*2, 8*). The following discussion highlights those parts of the theory that assist in understanding the parallel algorithm.

Consider the resonance energy between two HF type wave functions as in equation 2 where $\Psi^X = \left| \phi_i^X \phi_j^X \phi_k^X \ldots \right|$ is a normalized, antisymmetrized molecular wave function, and ϕ_i^X are the molecular orbitals (MO's). This problem is simplified by transforming the orbitals of Ψ^A and Ψ^B such that:

$$\left\langle \overline{\phi}_i^A \middle| \overline{\phi}_j^B \right\rangle = \lambda_{ij}^{A,B} \delta_{ij}$$

(3)

This biorthogonalization reduces the problem to the more standard-looking evaluation:

$$H_{AB} = \left\langle \overline{\Psi}^A \middle| \hat{H} \middle| \overline{\Psi}^B \right\rangle \text{ and } S_{AB} = \prod_{ij} \lambda_{ij}^{A,B}$$

(4)

where the overlap has been replaced by a product of the individual orbital overlaps. Expanding the above expression over molecular orbitals leads to:

$$H_{AB} = 2\sum_i \eta_i h_{ii}^{AB} + \sum_{ij} \eta_{ij} \left(2J_{ij}^{AB} - K_{ij}^{AB} \right) \quad \eta_i = \frac{S_{AB}}{\lambda_i}, \ \eta_{ij} = \frac{S_{AB}}{\lambda_i \lambda_j}$$

(5)

$$h_{ii}^{AB} = \left\langle \overline{\phi}_i^A(1) \middle| \hat{h} \middle| \overline{\phi}_i^B(1) \right\rangle \text{, the one electron term}$$

(6)

$$J_{ij}^{AB} = \left\langle \overline{\phi}_i^A(1)\overline{\phi}_j^A(2) \middle| \frac{1}{r_{12}} \middle| \overline{\phi}_i^B(1)\overline{\phi}_j^B(2) \right\rangle \text{, the coulomb term}$$

(7)

$$K_{ij}^{AB} = \left\langle \overline{\phi}_i^A(1)\overline{\phi}_j^A(2) \middle| \frac{1}{r_{12}} \middle| \overline{\phi}_j^B(1)\overline{\phi}_i^B(2) \right\rangle \text{, the exchange term}$$

(8)

Generalizing this result to the open-shell case requires treating the alpha and beta spin systems separately when performing the biorthogonalization. This treatment is necessary in order to produce transformations which leave the energy unchanged:

$$H_{AB} = \sum_{i\alpha} \eta_{i\alpha} h_{i\alpha,i\alpha}^{AB} + \sum_{i\beta} \eta_{i\beta} h_{i\beta,i\beta}^{AB} + \sum_{i\alpha,j\alpha} \eta_{i\alpha,j\alpha} \left(J_{i\alpha,j\alpha}^{AB} - K_{i\alpha,j\alpha}^{AB} \right) +$$
$$\sum_{i\beta,j\beta} \eta_{i\beta,j\beta} \left(J_{i\beta,j\beta}^{AB} - K_{i\beta,j\beta}^{AB} \right) + \sum_{i\alpha,j\beta} \eta_{i\alpha,j\beta} \left(J_{i\alpha,j\beta}^{AB} \right)$$

(9)

where the α and β indices indicate spin, and the η's are as defined in equation 5. Further generalizing this result for multi-determinantal wave functions:

$$\Phi^A = \sum_a C^{Aa} \Psi^{Aa} \qquad \Phi^B = \sum_a C^{Ba} \Psi^{Ba}$$

and   (10)

the matrix element can be rewritten, giving a sum of single-determinant pair energy calculations:

$$H_{AB} = \sum_a \sum_b C^{Aa} C^{Bb} \left\langle \overline{\Psi}^{Aa} \left| \hat{H} \right| \overline{\Psi}^{Bb} \right\rangle$$

(11)

Using a basis set expansion:

$$\overline{\phi}_i^{Aa} = \sum_\mu c_{\mu i}^{Aa} \chi_\mu$$

(12)

and rewriting H_{AB} in terms of density matrices, we have the following expression:

$$\sum_{ab} C^{Aa} C^{Bb} \left\langle \overline{\Psi}^{Aa} \left| \hat{H} \right| \overline{\Psi}^{Bb} \right\rangle =$$
$$\sum_{ab} C^{Aa} C^{Bb} \sum_{\mu\nu\lambda\sigma} \underbrace{D_{\mu\nu}^{ab} \left(T_{\mu\nu} + V_{\mu\nu} \right)}_{\text{one electron contribution}} + \underbrace{\left\langle \chi_\mu \chi_\nu \middle| \chi_\lambda \chi_\sigma \right\rangle \left(2 D_{\mu\nu}^{ab} D_{\lambda\sigma}^{ab} - D_{\mu\sigma}^{ab} D_{\lambda\nu}^{ab} \right)}_{\text{two electron contribution}}$$

(13)

where $D_{\mu\nu}^{ab} = \sum_i c_{\mu i}^{Aa} c_{\nu i}^{Bb}$ is a $\mu\nu^{th}$ pseudo-density matrix element for the ab^{th} determinant pair; $\left\langle \chi_\mu \chi_\nu \middle| \chi_\lambda \chi_\sigma \right\rangle$ is the two electron integral over basis functions; $T_{\mu\nu}$ is the kinetic energy over basis functions; $V_{\mu\nu}$ is the potential energy over basis functions; and we have incorporated the η's into the density matrices. The most time consuming part of the above calculation is the two-electron contribution, and our algorithm is dedicated to calculating this contribution efficiently.

Algorithm

Generally, the two electron integrals are stored as a set of all $\lambda\sigma$ indices for a specific $\mu\nu$ index. In the following discussion $\mu\nu$ and ab are "pair indices": they span all pairs of μ,ν (basis functions) and a,b (density matrices) indices, respectively. The notation $\mu\nu+1$ indicates that the next pair index (the usual sequence is 1,1; 2,1; 2,2; 3,1;...etc.), and in a similar manner $ab+1$ denotes the next density matrix pair index. In order to calculate the contribution for this particular set of integrals (pair index $\mu\nu$) and a particular density matrix (pair index ab), the following operations are necessary:

1) Read in or compute the integrals (all $\lambda\sigma$ for a particular $\mu\nu$ index)
2) Read in or compute the pseudo-density matrices (index ab)
3) Calculate the energy contribution for the $ab\mu\nu$ indices ($E_{ab\mu\nu}$)

Pipeline Algorithm The above operations are not independent: the two-electron energy calculation requires prior setup of the integral and density matrix information. Despite these interdependencies, the separation of the computation into the above operations represents the first opportunity for parallel computation. Each operation will have its own processor pool; in order to keep each pool simultaneously active, we use a pipeline to control the data flow: while a two-electron energy component, $E_{ab\mu\nu}$, is being calculated by one of the processor pools, the integral set $\mu\nu+1$ and density matrix $ab+1$ are being simultaneously read/calculated by processor pools two and three. In Figure 1, each labeled operation occurs simultaneously on different processor pools. When all three tasks are finished for the specified pair indices, the results flow as indicated and processing starts on the next set of pair indices.

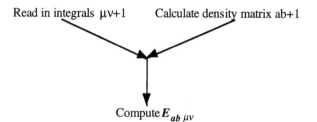

Figure 1: Pipeline for Computation of $E_{ab\mu\nu}$

The next obvious opportunity for parallelism would be to create multiple pipelines feeding into different portions of the energy calculation (Figure 2):

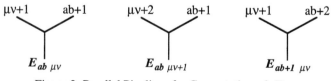

Figure 2: Parallel Pipelines for Computation of $E_{ab\mu\nu}$

However, this simple replication of the pipeline leads to inefficiencies: different pipelines recalculate identical setup information (integrals, density matrices) needed by other pipelines: in the above example the integral set $\mu\nu+1$ and density matrix $ab+1$ are each being setup twice. In order to avoid redundant computations, we reorganize the algorithm in terms of a grid of energy computations. Each block on the grid represents a single two-electron computation, $E_{ab\mu\nu}$, and a block's horizontal and vertical locations identify the prerequisite integrals and density matrices. A complete H_{AB} calculation requires traversal of the entire grid. The key to efficiently using integral and density matrix information lies in determining how to traverse the computational grid map, and how many grid locations to compute in parallel.

Truncated Wavefront   The solution we have chosen sweeps a "wavefront" of two-electron energy computations across the grid (Figure 3). At a particular step, the energy calculations performed are $(ab, \mu\nu-5)$, $(ab-1, \mu\nu-4)$, $(ab-2, \mu\nu-3)$, ..., $(ab-5, \mu\nu)$, where the pair indices correspond to the prerequisite density matrices and integrals, respectively. Concurrently, the $ab+1$ and $\mu\nu+1$ setup operations are performed. The energy calculations are represented graphically by a diagonal wavefront running from $(ab, \mu\nu-5)$ to $(ab-5, \mu\nu)$.

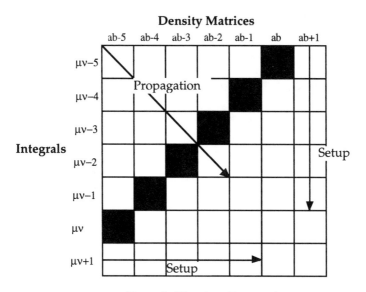

Figure 3: Wavefront Propagation

In the single pipeline (Figure 1) concurrent storage was required for only two sets of integrals ($\mu\nu$ and $\mu\nu+1$)and two sets of density matrices (ab and $ab+1$). Unfortunately, this is not the case for the wavefront scheme. As the wave progresses, all previous information (indices 0 to $ab+1$ or $\mu\nu+1$) is required; only when an axis has been completely swept is the setup information no longer necessary. This memory requirement is a serious problem since large systems may require more memory than is available. Truncating the wavefront is a simple solution. However, what is the most efficient way to propagate the truncated wave while still visiting every location in the calculation grid? One choice is propagating the wavefront along either of the two axes. For example (see Figure 4), if we restrict the range of integral indices to sets of three, we can propagate a wave of constant length three along the density matrix axis. Completing the first sweep across the density matrices results in the following contribution to H_{AB}:

$$\sum_{\mu\nu=1}^{3}\sum_{ab}E_{ab\mu\nu} \tag{13}$$

Truncating the calculational wavefront has many benefits. In our example the same integrals are used for the entire sweep along the density axis resulting in significant reuse of memory. Additionally, in a one-to-one mapping of the two-electron calculations to processors, the propagation described requires no movement of the integral buffers from processor to processor. As shown in Figure 4, processor one has integral set one stored in its local cache for the entire propagation of the wave along the density matrix axis. The same is true with processors and integral sets two and three. Wavefront truncation also allows a degree of freedom for optimizing the calculation: we can isolate the slowest setup step (reading the integrals in our example) and then perform this step least often by propagating along the *other* axis (the density matrices).

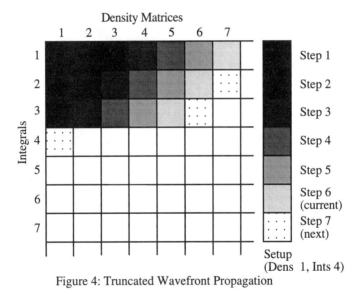

Figure 4: Truncated Wavefront Propagation

The wave formalism allows a smooth transition as one sweep ends and another begins. As shown in Figure 2, the seventh step, which would extend beyond the density matrix indices, "wraps around", restarting at density matrix 1. The obvious alternative to the wave formalism, a "vertical scan", does not allow for this smooth transition. The vertical scan would calculate $E_{ab\mu\nu}$ for $\mu\nu = 1$ to 3, $ab = 1$ while setting up $ab = 2$, and so on. When calculating $E_{ab\mu\nu}$ for $\mu\nu = 1$ to 3, $ab = 7$ (the final calculation of the row), the next setup required would be $\mu\nu = 4$ to 6 and $ab = 1$, which would result in a less efficient use of memory (storage for six integral buffers required rather than four) and more setup operations to perform, possibly reducing the number of processors in the calculation pool.

Load Balancing For the algorithm to run efficiently, we would like to minimize inter-processor communication and make continuous, non-redundant use of all available processors (load balancing). Even though inter-processor communication is implicit (*i.e.*, data moves upon memory access) on a shared memory multi-processor, it is still costly. We have already mentioned how either integrals or density matrices can be reused by choosing the wavefront direction. Additionally, the choice of wavefront direction also assists in load balancing.

However, more flexibility for balancing between the three processor pools is gained by performing a *composite* of energy component calculations at each grid location, rather than a single energy calculation. Specifically, we group the single energy calculations into larger, square blocks, with a length of **B**, to be determined later; these blocks are then used as the basic unit for the truncated wavefront (Figure 5). Although there will not necessarily be a one-to-one mapping of blocks to processors in most calculations, the arguments presented in favor of this organization still hold (minimal inter-processor communication and load-balancing).

Density Matrices

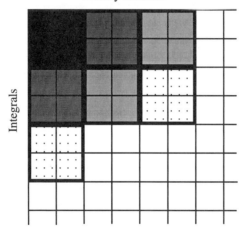

Figure 5: Truncated Wavefront Propagation, Using Blocking

The block length, **B**, is another degree of freedom which allows load balancing between the processor pools. The number of energy calculations per time step varies quadratically with **B**, since each block represents B^2 calculations. In contrast the number of setup operations varies linearly, since there are only **B** per block. Thus, by varying the blocklength, the algorithm can regulate the ratio of energy calculations to setup operations performed for each timestep. This regulation will ultimately be based on live timings of each sub-task.

The desire for no idle processor pools requires that the setup time per time-step is equal to the calculation time per processor per time-step; thus, each processor will complete its task at the same time as all others. One (or sometimes two) processors are dedicated to the setup group. To determine the two-electron calculation time per processor an expression for the total two-electron calculation time is required. The total number of integrals and density matrices stored, **M**, (assumed to be the same) is determined by the size of the available internal memory. **M/B** integrals and **M/B** density matrices can thus be stored (where B is the blocklength). Recalling that one group in memory is dedicated to the setup, we find $(M/B) - 1$ active blocks, giving:

$$B^2 * \left(\frac{M}{B} - 1 \right) = B(M - 1) \tag{14}$$

two-electron calculations per timestep. If the number of processors on the machine is **nproc**, the following is true:

$$\text{Setup time} = \frac{B(M - B) * \text{Two} - \text{electron calculation time}}{nproc - 1} \tag{15}$$

as one processor is devoted to the setup. Equation (15) is then solved for B. Ultimately, the algorithm will use live timing data for dynamic load-balancing; currently the timings are approximated based on machine specifications.

It is also important to know how the complete calculation will scale as we vary the number of processors. To do this, it is necessary to find an expression for the total calculation time. We know the ideal time for each step, the expression given above, and need only to determine the total number of steps a complete calculation requires. If there are T calculations to perform for a complete calculation, and $B(M-B)$ calculations per step, there are $T / (B(M-B))$ steps for a complete calculation. Thus, the total calculation time is

$$\frac{T * \text{Two} - \text{electron calculation time}}{nproc - 1} \tag{16}$$

and we find that the total calculation time should scale inversely with the number of processors. This expression is somewhat unusual since there is no explicit dependency on the setup operations; there is only an implicit dependency in the decrease of the number of processors available for the calculation. As the number of processors increases, this factor should be negligible, allowing optimal parallelization.

Program Architecture

The following describes the important internal characteristics of our program. We introduce here some of the software techniques that have proven useful.

C++ Programming Language Although far from being a standard in computational quantum chemistry, C++ allows easy organization of the program due to a natural mapping between standard chemistry concepts and computer code. The major data structures in our program are implemented as C++ classes or "objects". Objects are an encapsulation of *data* with associated *algorithms* or methods for manipulating that data. Many chemical concepts can be described naturally as objects and hierarchies of objects. For example, molecules can be thought of as data (atoms, basis sets, atomic coordinates), plus algorithms (basis set manipulations, coordinate transformations); in a likewise manner the component atoms in the molecule can also be described as objects. A high-level object in our program is the ResCalc object, which is used to organize the data (molecules, wave functions) and implement the algorithms (biorthogonalization, two electron energy contribution) needed for the resonance calculation. The natural mapping between chemical and program objects clarifies logic of the program and the structure of the data.

The Tcl Interpreter When designing the user-interface, it is desirable to maximize the amount of information a user can extract from the program. Additionally, the user should have a significant level of control over each important algorithm within the program. To accomplish this flexibility each major C++ object was provided with an interface to the command language that drives the program, the Tool Command Language (Tcl) (*9*).

Tcl is an *embeddable* and *extensible* command interpreter; it is embeddable because the interpreter is linked into our program and extensible because the native command set of the interpreter can be augmented by C/C++ code. Tcl provides a mechanism by which a text stream is interpreted while our code provides the implementation necessary for the commands. The Tcl language is simple, yet powerful enough to surpass the capabilities of most specialized "macro" languages used in computational chemistry codes. Tcl includes loops, conditional expressions, and variables. The main loop of our program consists of collecting input characters (either from a script file, a TCP/IP socket connection, or an interactive command line) and passing them into the Tcl interpreter.

Tcl Enabled Objects One of the most important Tcl commands that we have implemented is "new"; this command allows users to instantiate one of the chemical C++ objects. For example, to load a molecule, the user enters the command *new Molecule <molecule name>*. Once instantiated, a new chemical object provides additional Tcl commands having a nearly one-to-one correspondence with the methods provided by the underlying object. In the molecule example, there are commands for loading the molecular structure, loading wave function coefficients, extracting basis set overlap matrices, etc. We refer to these objects as *Tcl-enabled objects*, as they are C++ objects available at the user-level.

As a result of Tcl-enabling all important C++ objects, the user can instantiate, access, and control all of the major data structures and algorithms in the program. This enabling allows an unprecedented flexibility in constructing a calculation and inspecting its results. A few examples illustrate this point. Suppose the user wishes to construct an electronic state from a superposition of two wave functions, whose coefficients are stored on disk, and use the resulting wave function in a resonance calculation. Using the commands available to the Tcl-enabled Molecule object, the two wave functions can be loaded; their coefficients extracted to Tcl-enabled Matrix objects; these Matrices manipulated using standard linear algebra techniques; and the resulting Matrix returned back to a Molecule object. The modified Molecule object is then used in the resonance calculation. This is accomplished by the user without modifications or additions to the program.

As another example, consider the task of reading a wave function from a source not currently supported by the resonance program. For most programs this would require additional code to be linked into the program to support the new file format. However, using Tcl-enabled objects and text processing capabilities built into standard Tcl, the user can write a script to import foreign file formats directly into the Tcl-enabled Molecule object without needing to recompile and relink the program.

The use of Tcl and Tcl-enabled C++ objects has proven extremely useful during normal use and debugging as well. Because of the high degree of access to internal data structures and algorithms, many tests could be performed at the script level during debugging. For example, at a point where wave functions should be biorthogonalized, a debugging script can easily extract the wave function coefficients to Tcl-enabled Matrix objects, compute the overlap matrix, and check for a diagonal matrix. We found debugging time to be greatly decreased by reducing the need to install diagnostic print statements and reducing the need for recompiling and relinking.

Results and Discussion

All timing results reported come from calculations performed on a Silicon Graphics 4D/480, with eight 40 MHz R3000 processors. The system contains 256 Mb of shared internal memory. We chose two large problems of chemical interest to use for timing tests. The first, calculating the resonance energy of the cyclopentadienyl anion ($C_5H_5^-$, Cp), involves a five valence-state calculation where each state has three GVB-correlated pairs (perfect-pairing model) and each state localizes the negative charge on a different carbon atom. The second problem is the calculation of the resonance energy for the molecule 1,6-didehydro[10]annulene ($C_{10}H_6$), a ten membered ring that formally meets the Hückel criteria of (4n+2) π−electrons for aromaticity. The π−valence structure can be written as a resonance between two sets of five π−bonds:

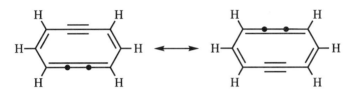

This system is of interest for understanding the activity of enediyne antitumor antibiotics (10). Both calculations were beyond the limits of older programs and thus presented fresh opportunities for our program.

The timing data presented represent the *elapsed* time to completion for each run. Ideally, this time should be related to the number of processors by the following relationship:

$$\text{total execution time on N processors} = \frac{\text{time on a single processor}}{\text{N (processors)}} \quad (17)$$

Figure 6 presents the timing results. As one of the processors is always devoted to a setup operation, the effective number of processors devoted to the parallel two electron energy calculation is nproc - 1 (equation 16). The x-axis represents the number of processors devoted to this parallel two-electron energy calculation, and the y-axis represents the speedup in total elapsed time. The scatter in the data is a result of running the program on an otherwise heavily loaded system. We were able to assure our program ran at a high priority by using *Nanny* (11); nevertheless, these background jobs still had a slight effect on our run times, probably due to I/O contention and extra context switching. Because of the other jobs on the system, and the fact that our runs could not have perfect utilization of the desired processors, we present in Figure 6 execution time speedups *scaled to 100% utilization of the desired processors*, to maintain constancy amongst the data.

Note that the Cp case begins to exhibit leveling off as the number of processors is increased, while the $C_{10}H_6$ case still exhibits linear speedups. The Cp case is much smaller than the $C_{10}H_6$ case, and we believe that its smaller size causes it to begin displaying non-linear speedups more rapidly than the larger $C_{10}H_6$ case. While this phenomenon is not desired, it is acceptable, as the larger cases are the ones for which greater speedups are needed.

Based on these results, we expect the speedups will scale well to larger numbers of processors. Recent results on distributed clusters of workstations support this projection(*12*). We are in the process of porting the program to a 64 processor KSR computer where additional tests can be made.

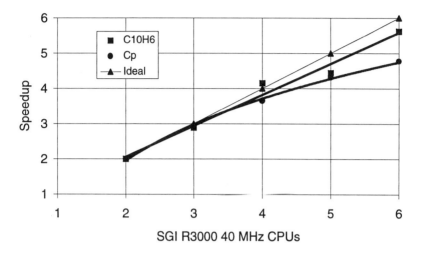

Figure 6: Effective Numbers of Processors vs.
Elapsed Time Speedups Scaled to 100% utilization

The values of H_{AB}, H_{AA}, and S_{AB} for Cp and $C_{10}H_6$ are shown in Table I (note that H_{AB}, H_{AA} are purely electronic energies; the nuclear repulsion energies are not included). For Cp the values are for the matrix elements between the specified charge localizations, either the 1-2 interactions (*ortho*) or the 1-3 interactions (*meta*). Solution of the secular equation for Cp leads to an E_{AB} of –192.185956986 hartree for the total wave function, compared to the single state energy of -192.17354519 hartree (both energies include nuclear repulsion terms).

Table I: Resonance Matrix Elements and Overlaps for $C_{10}H_6$ and Cp.

Resonance Interaction	H_{AB} (hartrees)	H_{AA} (hartrees)	S_{AB}
$C_{10}H_6$	–510.092102693	–796.224965290	0.640594
Cp-*ortho* charges	–330.798477408	–342.951275455	0.965792
Cp-*meta* charges	–336.133480369	–342.951275455	0.981394

Advances in *ab initio* techniques and computers have led to the efficient calculation of larger and larger HF and GVB descriptions of molecules. The ability to rapidly calculate resonance matrix elements for these large systems provides a way to study resonance and electron transfer problems with more rigor than previously feasible. The ability to use chemically intuitive basis states may help lead to a better understanding of the important energetics in electron transfer and other resonance-related problems.

Literature Cited:

1. Jackes, C. F.; Davidson, E. R. *J. Chem. Phys.*, **1976**, *64*, 2908.
2. Voter, A. F.; Goddard, W A., III *J. Chem. Phys.*, **1981**, *75*, 3638.
3. Unpublished programs, A. F. Voter, J. M. Langlois.
4. Bobrowicz, F. W.; Goddard, W. A., III in *Methods of Electronic Structure Theory*; Schaefer, H. F., III, Ed.; Modern Theoretical Chemistry, Vol. 3; Plenum Publishing Corp.: New York, NY, 1977; pp 79-127.
5. Goddard, W. A., III; Ladner, R. C. *J . Am. Chem. Soc.*, **1971**, *93*, 6750.
6. Ladner, R. C.; Goddard, W. A., III *J. Chem. Phys.*, **1969**, *51*, 1073.
7. Hunt, W. J.; Hay, P. J.; Goddard, W. A., III *J. Chem. Phys.*, **1972**, *57*, 738.
8. Voter, A. F. Thesis, California Institute of Technology, 1983.
9. Ousterhout, J. K., *Tcl and the Tk Toolkit*, Addison-Wesley: Reading MA, 1994.
10. Myers, A. G.; Finney, N. S. *J. Am. Chem. Soc.*, **1994**, *116*, 10986.
11. *Nanny* CPU-time balancer, Parallelograms, P.O. Box AA, Pasadena, CA 91102, info@pgrams.com.
12. Bierwagen, E. P.; Coley, T. R.; Goddard, W. A., III "*Ab Initio* Stuies of Electron-Transfer Rates Using Distributed, Parallel Computing", Talk presented at 1994 ACS meeting, Washington, D. C.

RECEIVED November 15, 1994

Chapter 8

Promises and Perils of Parallel Semiempirical Quantum Methods

Kim K. Baldridge

San Diego Supercomputer Center, P.O. Box 85608,
San Diego, CA 92186-9784

The application of semiempirical quantum mechanical procedures on Multiple-Instruction-Multiple-Data (MIMD) parallel computers is found to be a challenge. Key computations in these large scale quantum chemistry packages is the determination of eigenvalues and eigenvectors of real symmetric matrices. These computations arise in both geometry optimization as well as vibrational analyses, and, typically consume at least half (most, in the latter case) of the total computation time. This work illustrates the parallelization of both tasks within MOPAC. The application of the parallel code is demonstrated on several key molecular systems.

Utilization of computationally derived chemical and physical properties has vastly enhanced the success of experimental ventures into the creation of designer molecules of technological and medicinal importance. Rational drug design and novel nanomolecular materials would be complete fantasies if not for the atomic scale insight provided by computational chemistry. Because of the high demand for pharmaceuticals and composite materials to display a special uniqueness of action or efficiency in response, the tightness on specific structural tolerances and hence the degree of complexity in these molecular blueprints are increasing at a rate only manageable by advanced computing methods (e.g. massive parallelization, or ultrafast vectorization). Despite the extraordinary abilities of modern hardware technology and coding methods to manipulate the raw data, the rate limiting step in harmonizing the intricacy and precision required to push forward these chemical frontiers ultimately comes down to the optimization of the complex computational methodologies on state-of-the-art hardware platforms.

There are currently three commonly employed theoretical methods for the study of the properties of molecules: Molecular Mechanics; *Ab Initio*; Semiempirical. (Figure 1).

0097–6156/95/0592–0097$12.00/0

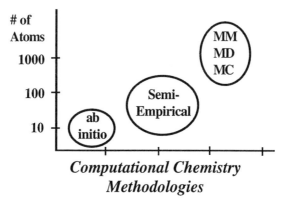

Computational Chemistry
Methodologies

Figure 1.   Schematic showing the relationship between theoretical method
and size of molecular system.

It has been well-established that quantum mechanical methods based on Hartree Fock
(HF) theory provide a successful and thoroughly tested framework for molecular
calculations (*1-2*). There are, however, major limitations in the size of molecular
systems that can be reasonably calculated on the available hardware. Computational
costs and complexity of solving the large iterative eigenvalue/eigenvector systems
associated with the theoretical methods become quite demanding (*3*). Even the fastest
computers have limitations on the size of molecular systems that can be solved due to
CPU time, memory, and disk space requirements. At present, the upper limit is about
1000 basis functions (Basis functions are mathematical functions which represent
atomic orbitals, as in descriptive organic chemistry. The number of basis functions
used in a calculation of a particular molecule determines the level of accuracy of that
calculation, and forms what is called a basis set), which corresponds to less than 40
first row atoms at a modest level basis set, i.e, about a tetrapeptide. On the other hand,
molecular mechanics and molecular dynamics techniques are extremely fast empirical
methodologies which are able to handle very large molecular systems, such as entire
enzymes with over 100 peptide residues. These methods sacrifice in generality and
accuracy. In addition, they are not parameterized for other than ground state systems,
and are unable to adequately represent geometries involved in bond-making/bond-
breaking processes.
        Between Hartree Fock methods and empirical-based methods are semiempirical
methods. Like *ab initio* methods, they are basically quantum mechanical in nature, the
main difference being that the semiempirical methods involve additional approximations
based on experimental data, thus simplifying the calculations considerably.
Semiempirical methods are right on the verge of becoming of routine use in polymer
and biochemical applications. The major constraint, despite the numerous
methodological advances in past years (*4-8*) is that the size of chemical systems that can
be analyzed, is largely a function of available single-processor computer power.
Although this power continues to increase in magnitude, it cannot continue to improve
at a rate that keeps pace with the desires and expectations of the scientific community.
Parallel architectures promise to make calculations of this size more of a reality.
However, only recently has it been realistic to turn towards the parallel computing
environment for any of these types of calculations (*9-12*) primarily due to the fact that
new distributed-memory algorithms that utilize the architectures of the parallel platforms
must be developed.
        This chapter focuses on the promises and concerns of applying parallel methods
to semiempirical calculations for the solution of problems that are currently not possible
with either *ab initio* or parallel *ab initio* methods (*13-18*) and with an accuracy greater

than that achievable with the molecular mechanics and dynamics type procedures. The conversion and performance evaluation of the semiempirical quantum chemistry code, MOPAC (*19-21*) on the Intel iPSC/860 and Paragon platforms will be demonstrated.

Semiempirical Quantum Methods

The primary goal of quantum chemical codes is to solve the molecular Schrodinger equation   (*22-25*). This involves the solution of the generalized eigensystem

$$A\ x_i = \lambda_i\ x_i$$

where **A** is a given n x n real symmetric matrix, and (λ_i, x_i) is one of n eigenvalue/eigenvector pairs to be determined. The solution of this eigensystem provides the molecular wave function, from which a total description of the molecule, including all molecular properties such as equilibrium geometry, dipole moments, energetics, kinetics, and dynamics is obtained. The applications programs (*26-28*) for these theories are typically large and complex, and large real symmetric eigenproblems (*29-35*) arise in various options, notably self-consistent field (SCF) (*36*) computations and molecular vibration analysis.

In SCF computations, **A** is typically the matrix representation of the Fock operator with respect to a given set of basis functions (atomic orbitals). The eigenvalue λ_i is an energy level corresponding to a molecular orbital represented as a linear combination of basis functions (atomic orbitals), with the components of the eigenvector x_i as the basis function coefficients. The matrix dimension n is the number of basis functions used in the computation, which varies roughly with the number of electrons in the molecule and the desired accuracy of the molecular orbital function representation. Values of n on the order of a few hundred are easily reached in even moderate-sized systems with several heavy atoms.

The SCF computation is iterative in nature, as the Fock operator depends on its own eigenfunctions, and the Fock matrix is usually constructed from the orbitals computed on the previous iteration. Thus, a sequence of eigensystems must be solved until convergence is attained. Moreover, the SCF iteration often is the inner iteration in a geometry optimization in which the nuclear coordinates are optimized with respect to energy. Thus, a single geometry optimization for a molecule with even a few heavy atoms (light atom refers to hydrogen; heavy atom refers to all other types) may require the solution of hundreds of large real symmetric eigensystems.

In vibrational analyses, the matrix **A** is the Hessian of the energy with respect to the 3*N - 6 (N = number of atoms in the molecule) degrees of vibrational freedom in the nuclear coordinates. The eigenpairs (λ_i, x_i) determine vibrational frequencies and corresponding normal modes. The vibrational eigensystems are usually dimensionally somewhat smaller than in the SCF case, but again they may need to be solved repeatedly, for example, as part of a reaction path following computation.

In *ab initio* SCF computations, the matrix element computations involve the evaluation of up to $O(n^4)$ floating point operations for the evaluation of Coulomb and Exchange (interaction) integrals, whereas the solution of a single eigensystem is $O(n^3)$ (i.e. evaluation of the integrals dominate the computational effort). In semiempirical techniques, an approximate Hamiltonian is used so that the number of calculated Fock matrix elements is greatly reduced. These methods are based on the assumption that only electrons on the same atoms have significant interaction energies; all others are represented via experimental parameters. This reduces the calculation of integrals to

$O(n^2)$ and thus, solution of the eigensystem becomes the primary computational effort. MOPAC supports four semiempirical Hamiltonians: MNDO (*37*), MNDO/3 (*38*), AM1 (*14*), and PM3 (*39*). These are used in the electronic part of the calculation to obtain molecular structures, molecular orbitals, heats of formation, and vibrational modes. The advantages of semiempirical over *ab initio* methods are that semiempirical methods are several orders of magnitude faster, and thus calculations for larger molecular systems are possible by using one of these semiempirical Hamiltonians. The reliability of these methods in predicting accurate geometries and heats of formation has been demonstrated in many applications (*40-42*)

Parallel MOPAC: Structure and Task Distribution

MOPAC is public-domain software available through QCPE (*43*). Version 6.0 of MOPAC runs on VAX, CRAY and workstation platforms, and consists of approximately 50,000 lines of FORTRAN code, in 190 subroutines. Resident memory usage in MOPAC is governed entirely by parameter settings chosen at compile time. The amount of storage required by MOPAC depends on the number of heavy (non-hydrogen) and light (hydrogen) atoms that the code has been parameterized to handle at compile time, and whether configuration interaction capabilities are incorporated.

Hardware performance monitoring (*44*) (HPM) indicates that the majority of the computational time required to run MOPAC is divided among evaluating the electronic interaction integrals (Hartree Fock matrix preparation), calculating first derivatives (geometry optimization procedure), calculating second derivatives (vibrational analysis) and solving the resulting eigensystem (diagonalization). (Figure 2).

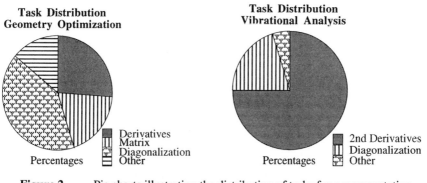

Figure 2. Pie charts illustrating the distribution of tasks for a representative geomery optimization and vibrational analysis calculation.

The precise division of CPU time among the tasks for a geometry optimization procedure may vary with molecular composition (Timings indicate that diagonalization can represent from 40-80% of the total computational load, depending on molecular construction); however, the general procedures which dominate the work load for the total calculation will remain the same. In general, semiempirical methods process N^2 integrals instead of N^4 as with conventional HF methods, therefore, the computational bottleneck lies at the diagonalization routine, giving an overall N^3 time dependence. The SCF calculation, geometry optimization, and second derivative evaluation (vibrational analysis) for the available Hamiltonians were parallelized in this work.

The parallelized algorithms were implemented on a 64-processor Intel iPSC/860 hypercube, and subsequently on an Intel Paragon; both distributed-memory, message-passing parallel computers. In the Intel hypercube, each processing node contains an Intel i860 CPU and 8 Mbytes of RAM (16 and 32 Mbytes/node on the Paragon). The communication links are through Intel's Direct-Connect Communications (DCM) hardware with a 2.8 Mbyte/sec maximum bandwidth for the iPSC/860 and 10-12 Mbytes/sec on the Paragon. (The band width on the Intel Paragon is widely variant depending on the system configuration; potentially, one could see a value as high as 4-5 times this.)

Geometry Optimization Component

As prompted by profiling techniques, detailed inspection of the algorithmic format of MOPAC shows that most of the computational work in the semiempirical geometry optimization procedure is distributed over the following three tasks:

1) Evaluation of one- and two-electron matrix elements.
2) Formation of the Fock matrix and diagonalization.
3) Evaluation of derivatives.

Sequential bottlenecks are the limiting case of poor load balancing. Of primary concern in choosing a parallel scheme is to ensure that some processors are not sitting idle awaiting results of others. Three basic techniques were considered in the parallelization of MOPAC (Table I): domain decomposition, control decomposition and statistical decomposition. In domain decomposition, the domain, or data, to be dealt with is a set of rectangles, n in number. Since every rectangle is the same amount of work, we send n/p rectangles to each of p processors. Theoretically, nearly perfect load balancing can be obtained using this method.

More often, the data can not be easily split into neat even packets of work. In cases like these, one can employ control decomposition in which case a formula is devised which approximately balances the work load across processors, based on the type of computation that is being performed. Finally, statistical decomposition is a parallel strategy for programs where the work load is dependent on the complexity of the problem the user has specified at run time. In the specific case of quantum codes, the work load involved in the calculation of the various types of integrals would, in many cases, benefit from a statistical parallel decomposition. The following section elaborates on the application of these decomposition schemes to MOPAC.

Table I. Parallel Decompositions

Decomposition Scheme	Characteristic
Domain	Domain (data) is a set of n rectangles, distributed over p processors.
Control	Domain is distributed in uneven packets to p processors.
Statistical	Domain is distributed according to complexity of run-time problem.

1) <u>Evaluation of one- and two-electron elements.</u> Geometry optimization begins with a call to a controller for the specific optimization method. This routine makes several calls to subprograms to carry out the various aspects of full optimization. Much of the calculation occurs in setting up the Hamiltonian matrix (Scheme I: Task 1, Loop over ATOMS). The resulting matrix elements are used to calculate the SCF heats of formation, the nuclear energy, and the one- and two-electron interaction integrals. MOPAC is based on a semiempirical approach, therefore, many of the integrals are ignored, others are calculated using experimental parameters stored in common blocks, and a few are calculated fully.

The computation of the one-electron and two-electron integrals has been distributed over nodes by partitioning the number of atoms over nodes and giving each node an independent number of integrals to calculate. In general, 100 integrals are calculated for each heavy atom - heavy atom interaction, 10 integrals for each heavy atom - light atom interaction, and 1 integral for each light atom - light atom interaction. Ideal load balancing can be achieved by splitting up the integrals in accord with the type of interaction so that each node receives approximately equal work to do, i.e. statistical decomposition. Because each two-electron integral contributes to several Fock matrix elements, it is necessary to have the independent node results collected before the Fock matrix is created. A way around this is to have each processor work on its own partial Fock matrix, which is gathered once at the very end. The construction of MOPAC makes this more difficult, but is currently being investigated.

2) <u>Formation of Fock matrix and Diagonalization.</u> The formation of the Fock matrix involves computation of the remaining contributions to the one-center integrals, and the two-electron two-center repulsion terms. Each of these subtasks is split over nodes in accord with the number of atoms (Scheme I: Task 2a, Loop over ATOMS). Once this is done, the density matrix can be computed along with information about orbital occupancy. This task is distributed over nodes in accord with the number of orbitals (Scheme I: Task 2b, Loop over ORBITALS).

MOPAC employs a combination of techniques for complete diagonalization. A "fast" or pseudo-Jacobi diagonalization procedure is invoked in initial SCF iterations. The diagonalizations during the final SCF iterations are then taken over by a more rigorous QL algorithm (*45-49*).

Typically, a diagonalization method consists of a sequence of orthogonal similarity transformations. Each transformation is designed to annihilate one or more of the off-diagonal matrix elements. In the case of the Jacobi method, successive transformations then undo previously set zeros, but the off-diagonal elements continue to decrease until the matrix is diagonal to the precision of the machine. Accumulating the product of eigenvector transformations gives the matrix of eigenvectors, and the elements of the final diagonal matrix are the eigenvalues. In general, the QL (QR if the matrix is reversed graded) algorithms are much faster than the Jacobi methods, however, the Jacobi methods can be computationally time-favorable relative to QL if a good initial approximation is available, and only a single Jacobi-sweep is done.

MOPAC replaces the full QL eigensolution by a single Jacobi-like sweep of just the occupied-virtual block for intermediate SCF iterations often with considerable speed enhancements (*50*). The algorithm is considered a pseudo-*diagonalization* technique because the vectors generated by it are more nearly able to block-diagonalize the Fock matrix over molecular orbitals than the starting vectors. It is considered *pseudo* for several reasons (*3*), the most important of which is that the procedure does not generate eigenvectors. In the chemical sense, the full orbital matrix representation is not diagonalized, only the occupied-virtual intersection is. All of the approximations used in this pseudo-diagonalization routine become valid at self-consistency, and further, the approach to self consistency is not slowed down (*51*).

Given the lower half triangle of the matrix to be diagonalized in packed form, the algorithm has three primary loop sequences that constitute the procedure (Scheme I: Task 2c, Loop over VARIABLES). The first two loops together perform the similarity transformation

$$V^t F V$$

that transforms from the atomic orbital to the molecular orbital representation (F represents the Fock matrix). This representation ensures that the resulting eigenvectors are orthogonal, spanning the N-atom dimensional space. This step is followed by rotation, which eliminates off-diagonal elements. The matrix is then block diagonalized only, because only Fock elements connecting occupied and virtual orbitals must be zero at convergence.

Two methods of parallel decomposition were investigated for the diagonalization procedure. The initial attempts distributed the work load over rows or columns of the matrix, i.e., control decomposition. This method resulted in timings that were actually significantly slower than the original unparallelized routine. This is due to large communication overhead from processing such small amounts of data. In addition, two utility routines were written to establish each node's starting work load position. Calls to these routines, along with additional global calls to gather and broadcasts to announce individual node data, resulted in extreme overhead costs.

To avoid some of the complications of the above, a domain decomposition was employed. In this method, large groups or blocks of the matrix are distributed over nodes. Parallelization in this manner eliminates the need for broadcasting intermediate results. Only the final computed vectors are gathered via a global routine. Broadcasting of intermediate results is no longer necessary and scratch arrays already available are used for parallel decomposition so that no additional memory is required for this parallel method.

3) Evaluation of derivatives. Additional CPU-intensive subroutines involved in the geometry optimization include those that carry out derivative evaluation (Scheme I: Task 3). The derivatives of the energy with respect to the internal coordinates is done via finite differences. The total work involves $3*N$ variables that can be distributed equally over the number of nodes.

Vibrational Analysis Component

Vibrational analysis (second-derivative evaluation) of molecular systems can be a formidable task. These calculations are, however, essential to characterize stationary points and to assess vibrational and thermodynamic properties of molecules. The vibrational analysis procedure involves construction of a $3*N$ dimensional matrix of second derivatives of energy with respect to Cartesian coordinates (Scheme II). The calculation of each of these matrix elements represents an independent calculation, and the procedure holds the potential of being perfectly parallel. Following the calculation of matrix elements across nodes, the results are collected using a global routine and the full matrix diagonalized.

The diagonalization of the matrix results in a set of eigenvectors, corresponding to the $3*N-6$ vibrational motions, and a corresponding set of eigenvalues, which represent the respective vibrational frequencies of these motions. The other 6 eigenvectors correspond to the rotational and translational motion, with associated zero eigenvalues (disregarding numerical artifacts).

Scheme II shows the vibrational analysis procedure. The parallelization of the vibrational analysis component requires partitioning $3*N$ variables over nodes to calculate a matrix of second derivative elements. Because this is a symmetric matrix, there are $3*N*(3*N+1)/2$ unique elements to be computed. It is critical to maintain proper indices over the nodes as the results are calculated. A global routine is invoked to collect the matrix in preparation for diagonalization.

SCHEME I: Parallelization of Geometry Optimization

Specific Task	Loop Sequence

Specific   **Loop Sequence**
Task

        Loop according to SCF criteria (**COMPFG**)
1   ==>Evaluation of Hamiltonian matrix elements (**HCORE**)
        Loop over total number of ATOMS
          * fill 1 e- diagonal/off diagonal of same atom
          * fill atom-other atom 1 e⁻ matrix (**H1ELEC**)
          * Calculate 2 e⁻ integrals
        Calculate e⁻ - nuclear terms   (**ROTATE**)
        Calculate nuclear-nuclear terms
        Merge 1 electron contributions private to each CPU
2   ==>Formation of Fock matrix and Diagonalization (**ITER**)
2a   Loop over number of ATOMS
          * remaining 1 e- elements (**FOCK1**)
          * 2 e-/2-center repulsion elements of Fock matrix (**FOCK2**)
2b   Loop over number of ORBITALS
          * density matrix (**DENSIT**)
2c   Loop over matrix BLOCKS
        Diagonalization (**DIAG**)
          * Construct part of the secular determinant over MO's
            which connect soccupied & virtual sets.
          * Crude 2x2 rotation to "eliminate" significant elements.
          * Rotation of pseudo-eigenvectors.
        Merge contributions private to each CPU
3   ==>Evaluation of Derivatives (**DERIV**)
        Loop over number of ATOMS
         * derivatives of energy w.r.t. Cartesians (**DCART**)
        Loop over number of VARIABLES
         * Jacobian : d(Cart)/d(internals) (**JCARIN**)
        Merge derivatives private to each CPU

SCHEME II: Parallelization of Vibrational Analysis

        Calculation of force constants and vibraional frequencies (**FORCE**)
        Loop over number of VARIABLES
         * Calculate second-order of the energy with repect to
            the Cartesian coordinates (**FMAT**)
        Merge second derivative components private to each CPU

Results

Code performance was demonstrated on a large group of molecules (*52*) varying in symmetry construction and heavy atom/light atom ratios; only a few of these molecules are presented here to illustrate the general trends. Although a principal performance measure is the elapsed time necessary to solve the problem of interest, speedup shows more clearly the behavior of a parallel program as a function of the number of processors. Speedup is defined as the wall-clock time to run on 1 processor, divided by the time to run on p processors. For a perfectly parallelized code, speedup equals the number of processors. Single processor timings are taken as the best serial algorithm (In this case, timings for serial and parallel MOPAC on one node are identical, due to the particular method of parallelization.).

The parallel procedures were first implemented on the Intel iPSC/860, and subsequently on the Intel Paragon, at the San Diego Supercomputer Center. Since the iPSC/860 had only 8 Mbyte nodes, the molecular constitution was limited to less than 20 heavy atoms. In general, code performance is identical on both platforms with the exception of a 25% faster clock in the Paragon, thus shifting the resulting curves by the appropriate amount. The geometry optimization and vibrational analysis computations are illustrated for corannulene, which has 20 carbon and 10 hydrogen atoms (Figures 3 and 4). One notices from these curves that, because of the faster clock in the Paragon, the workload per node is done faster, but the general trends are virtually identical.

All molecules investigated serve as prototypes for the classes of molecules that we intend to study computationally for these types of methods. These include prototypes for aromatic carbon materials based on graphitic or fullerene motifs, prototypes for strained polycyclic hydrocarbon-based "energetic" materials, and prototypes for pharmacophores and bioreceptor substrates. For detailed discussion of Paragon performance, three molecules from the total set were chosen: norbornyne cyclotrimer, taxol derivative, and lophotoxin (Figure 5).

Calculations were performed on node combinations up to approximately 128 32 Mbyte processors. Optimization level 3, which incorporates global optimization and software pipelining, was invoked during code compilation. Speedups approaching 5 and absolute Intel speeds of about half that of the CRAY C90 were obtained. Figure 6 shows a plot of CPU time versus number of processors for these three molecules. One finds a definite compartmentalization of data with respect to optimal number of processors for specific number of atoms in a molecular system. This graph shows an optimal performance of 64 nodes for the molecules considered, which range from 45-60 atoms. Scanning the entire data base of molecules (*20*) shows the general rule of thumb.

Number of atoms, n	Number of nodes at optimal performance
> 29	64
15 < n < 30	32
9 < n < 16	16
< 10	8

Running the calculation on more than this optimal number of nodes will be inefficient due to either a very small distribution of work across nodes, or some nodes being left completely idle. In some cases, performance is degraded due to the fact that the communications costs start to dominate due to many nodes transferring very small amounts of data. This is clearly seen in the performance plots of corannulene from the Paragon.

Although the overall speedup appears inefficient for geometry optimization, individual task speedups illustrate the promise of parallelization. Figures 7 and 8 show the breakdown over the three tasks parallelized for lophotoxin and the taxol derivative,

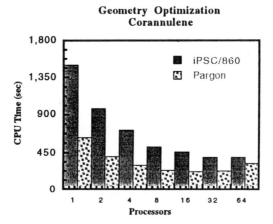

Figure 3.   Bar chart comparing the performance of the iPSC/860 and
Paragon or the geometry optimization of corannulene.

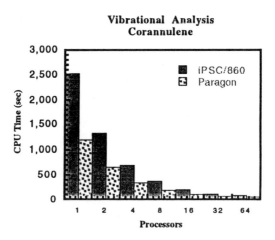

Figure 4.   Bar chart comparing the performance of the iPSC/860 and
Paragon for the vibrational analysis of corannulene.

Corannulene

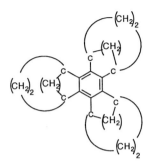

Norbornyne cyclotrimer

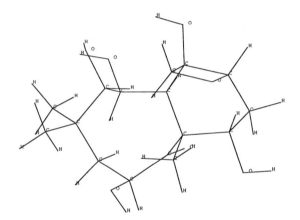

Taxol derivative

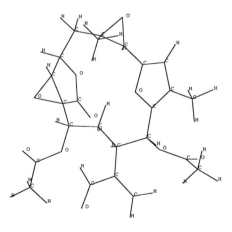

Lophotoxin

Figure 5. Molecular structures.

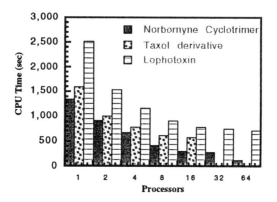

Figure 6.   Speedup curve illustrating the variation in speedup with number of
processors for a geometry optimization calculation for norbornyne
cyclotrimer, taxol derivative and lophotoxin.

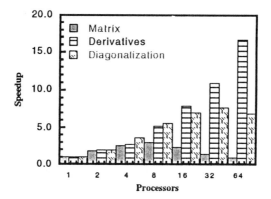

Figure 7.   Bar graph showing the effects of parallelization of the various tasks
involved in the geometry optimization calculation for the taxol
derivative.

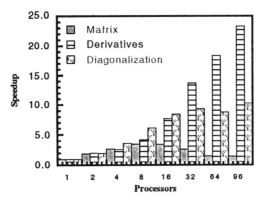

Figure 8.   Bar graph showing the effects of parallelization of the various tasks
involved in the geometry optimization calculation for lophotoxin.

respectively. The parallelization of the derivative routine is the most impressive, approaching a speedup factor of 25, while that of diagonalization and formation of matrix elements lag around 10 and 5, respectively. Geometry optimizations that involve a diagonalization task that is closer to 80% of the total time will obviously give much more impressive results. Notice especially, the decay of the calculation of the matrix elements after about 8 nodes. This task is generally the smallest effort of the three tasks, and, for the size of molecules considered here, does not have enough of a workload to keep more than about 8 nodes busy, before communication costs supersede work time.

Timing results for the parallelized vibrational analysis procedure are very encouraging. For all molecules considered, the CRAY speed is in the range 16-32 Mflops. The Intel results on the other hand approach or, in the case of lophotoxin, exceed 200 Mflops. The vibrational calculations show nearly linear speedups for all sizes of molecules. As the work is distributed over more and more nodes, efficiency is lost, especially noted for very small molecules. One factor contributing to this loss is that there is less and less work to distribute over nodes. This is why the larger molecules show better performance than the smaller molecules. In addition, a latency effect could be contributing to a decrease in efficiency due to more nodes being involved in the global calls; this effect is uniform over all sized molecules. The speedups for the three molecules considered here clearly show a linear trend (Figure 9). The results are fairly uniform for molecules of similar size, because the primary task in any particular molecule is the calculation of 3N (N=#atoms) second derivatives of energy with respect to coordinates.

An important issue here is the range of problem sizes for which the performance is acceptable. Keeping the number of processors fixed and increasing the problem size increases the amount of local computation each node does, therefore, performance is expected to improve for larger molecular systems. This is illustrated in Figures 10 and 11 for 64 node results over the entire range of molecules for geometry optimization and vibrational analysis, respectively. Similar curves are obtained for the other node combinations.

Discussion

A major limitation on performance, particularly for the geometry optimization calculations, is the code memory requirement. Even for the largest molecules calculated, there is a noticeable asymptote in the speedup curve as the number of nodes increases. This is primarily because the molecular systems are relatively small in comparison to the number of nodes being allocated to do work, a restriction resulting from the memory constraints. The main problem in MOPAC stems from the rather poor structure of the code in terms of memory utilization. The use of replicated data parallel decomposition requires sufficient memory be held on each processor for the entire symmetric Hamiltonian and Fock matrix. As a result, all internal communications throughout parallelized MOPAC are carried out with fast-library global routines and not via sending/receiving packets of information. This method of parallelization was chosen in order to minimize the communication overhead and latency costs, which were observed to be extremely high, especially with the first levels of operating systems on the MIMD machines. There will still be startup time for these global routines that will contribute to the overall time costs, however, this is much less due to better algorithmic construction with global routines, and the fact that the global routines are faster than the send/receive routines.

A second limitation, as noted for large biomolecular systems, is the instability of the geometry optimization algorithms. If one tries to calculate the structure of a large, floppy biomolecular system (i.e, > 100 atoms), there is a serious problem with convergence due to the many torsional degrees of freedom. The semiempirical algorithms that are currently available are not sensitive enough to instigate convergence.

Unfortunately, due to the severe memory and algorithmic constraints, the goal of being able to calculate larger molecules than can be currently calculated with *ab initio*

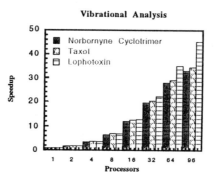

Figure 9. Speedup curve illustrating the variation in speedup with number of processors for a vibrational analysis calculation for nobornyne cyclotrimer, taxol derivative and lophotoxin.

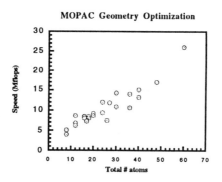

Figure 10 Plot showing the increase in Mflop rate with increase in size of molecular system for a geometry optimization calculation. The molecules range in size from 8 atoms to 60 atoms.

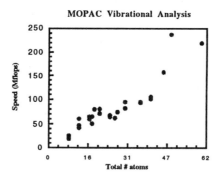

Figure 11. Plot showing the increase in Mflop rate with increase in size of vibrational analysis calculation. The molecules range in size from 8 atoms to 60 atoms.

techniques has not been met. The declining cost of semiconductor memory makes it reasonable to assume that large-scale parallel computers will provide sufficient memory per node to accommodate much larger molecules with the existing software. In addition, memory need not be all semiconductor memory; one could think of employing common file system disk storage to accommodate large intermediate information such as integrals. Still, one will inevitably reach the limits of the increased memory capabilities. Therefore, we are currently investigating both algorithmic modifications and distributed memory capabilities. Distributed memory can potentially be handled with system routines, which allocate appropriate memory at the onset of the problem, or else hard coded dynamic memory using large fixed arrays in common with pointer capabilities. Preliminary results by this author (current work involves MOPAC 7.0 on the Intel Paragon and T3D platforms) as well as others (*53*) in this area show much more promise towards the calculation of molecules of the size of hundreds of atoms.

Conclusions

With (MIMD) computers clearing the way for record-breaking computation speeds, scientific programmers of the 90s are being pushed to the world of parallel programming. Massively parallel processors achieve their high speed by working on many parts of the problem in parallel. While it is difficult in many cases to structure a problem for efficient highly parallel solution, for those problems for which the technique is applicable, these computers are an increasingly important computational tool, especially for large and difficult chemistry problems. Thus, it is clear that implementing chemistry applications in parallel environments is a milestone for computational chemistry.

In this work, we have demonstrated the promise as well as the difficulties involved in the implementation of semiempirical quantum chemistry applications on the Intel hypercube platforms. As the first level of implementation of these methods, we have employed a replicated data parallelism strategy. In this strategy, even though tasks are distributed over nodes, results of all distributed tasks are collected together on each node (replicated) at various points within the Hartree-Fock procedure, thereby causing limitations, especially for the geometry optimization calculation, due to the amounts of memory necessary to hold these quantities on each individual node. This severely limits the size molecular system that can be calculated and forces an unacceptably low ratio of processors to memory. With less than 32 Mbytes/node, the size of molecular systems that can be modeled is limited to less than 60 atoms, and the speedup saturates at 16 to 32 nodes.

The vibrational analysis component shows more promise within the replicated data parallel implementation. The fact that the parallel implementation of the code performs similarly for large and small systems, allows us to extrapolate the results to predict that very large numbers of processors could be brought to bear on this problem given efficient global calls and memory. Analogously, this work has shown the potential for parallelization of the vibrational component within *ab initio* codes.

Significant attention by this author as well as others (*53*) is now being given towards the implementation of these methods using a distributed data parallel strategy, which clearly shows to be superior in light of the known memory problems associated with these methods. In the distributed data parallel strategy, individual node tasks are not collected together on each node at any time during the calculation. Thus, performing a quantum mechanical calculation on a molecule of size N atoms, can be distributed over p processors such that only N/p amount of memory is ever needed on any individual node. This will allow our goals involving calculation of molecules with hundreds of atoms, and study of reaction paths and solvent effects of large systems to be a reality.

Acknowledgments

The author would like to thank Scott Kohn and Jay Siegel for several beneficial discussions. Support was provided by the National Science Foundation (Grant No. ASC-9212619 and Grant No. ASC-8902827), Intel Corporation, and the San Diego Supercomputer Center.

Literature Cited

1. Hehre, W. J.; Radom, L.; Schleyer, P. v. R.; Pople, J. A. *Ab Initio Molecular Orbital Theory;* John Wiley & Sons: New York, **1986**, and references therein.
2. Schaefer, H. F. III, *Science*, **1986**, *231*, 1100, and many examples cited therein.
3. Stewart, J. J. P.; Csaszar, P.; Pulay, P. *J. Comp. Chem.*, **1982**, *3*, 227.
4. Pulay, P. In *Applications of Electronic Structure Theory;* Schaefer, H.F., Ed., Plenum Press: New York, **1977**, p. 153.
5. Pople, J.A.; Krishnan, R.A.; Schlegel, H.B. *Int. J. Quantum Chem. Symp.*, **1979**, *13*, 225.
6. Morokuma, K.; Kato, S. In *Poitential Energy Surfaces andDynamics Calculations*; Truhlar, D.G., Ed., Plenum Press: New York, 1981, p. 243; Morokuma, K.; Kato, S.; Kitsura, K.; Ubara, S.; Ohta, K.; Hanamura, M. In *New Horizons in Quantum Chemistry*; Reidel: Dordrecht, **1983**, p. 221.
7. Morokuma, K.; Kato, S.; Kitaura, K.; Ubara, S.; Ohta, K.; Hanamura, M. In *New Horizons in Quantum Chemistry;* Lowdin, P.-O., Pullman, B., Eds., Reidel: Durdrecht, The Netherlands, **1983**, 221.
8. Gaw, J.F.; Yamaguchi, Y.; Schaefer, H.F. *J. Chem. Phys.*, **1984**, *81*, 6395.
9 Luethi, H. P.; Mertz, J. E.; Feyereisen, M. W.; Almlof, J. E. *J. Comp. Chem.*, **1992**, *13*, 160.
10. Whiteside, R. A.; Binkley, J. S.; Colvin, M.E.; Schaefer, H. F. III *J. Chem. Phys.*, **1987**, *86*, 2185.
11. Hertz, J. E.; Andzelm, J. W. CRAY Channels, **1991**, 10.
12. Luethi, H. P.; Mertz, J. E. Supercomputing Review, **1992**.
13. Whiteside, R. A.; Binkley, S. J.; Colvin, M.E.; Schaefer, H. F. *J. Chem. Phys.*, **1987**, *86*, 2185.
14. Guest, M. F.; Harrison, R. J.; van Lenthe, J. H.; van Corier, L. C. H. *Theoret. Chim. Acta*, **1987**, *71*, 117.
15. Dupuis, M.; Watts, J. D. *Theoret. Chim. Acta*, **1987**, *7*, 91.
16. Carbo, R.; Molino, L.; Calabuig, B. *J. Comp. Chem.*, **1992**, *13*, 155.
17. Luethi, H. P.; Mertz, J.E.; Feyereisen, M. W.; Almlof, J. E. *J. Comp. Chem.*, **1992**, *13*, 160.
18. Schmidt, M.W.; Baldridge, K.K.; Boatz, J.A.; Elbert, S.T.; Gordon, M.S.; Jensen, J.H.; Koseki, S.; Matsunaga, N.; Nguyen, K.A.; Su, S.; Windus, T.L., *J. Comp. Chem.*, **1993**, *14*, 1347.
19. Stewart, J. J. P. *Quantum Chem. Exchange. Bull.*, **1985**, *5*, 133, QCPE Program 455.
20. Dewar, M. J. S.; Zoebisch, E. G.; Healy, E. F.; Stewart, J. J. P. *J. Am. Chem. Soc.*, **1985**, *107*, 3902.
21. Stewart, J. J. P. *MOPAC Manual; A General Molecular Orbital Package,* Frank J. Seiler Research Laboratory, Dec. **1988**.
22. Levine, I. N. In *Quantum Chemistry*; Shull, Harrison, Ed.; Allyn and Bacon, Inc.: New York, **1983**.
23. Szabo, A.; Ostlund, N. In *Modern Quantum Chemistry;* MacMillan Publishing Co., Inc.: New York, **1982**.

24. Hehre, W. J.; Radom, L.; Schleyer, P. v. R.; Pople, J. A. In *Ab Initio Molecular Orbital Theory;* John Wiley and Sons: New York, **1986**.
25. Clark, T. In *A Handbook of Computational Chemistry;* John Wiley and Sons, Inc.: New York, **1985**.
26. Frisch, M. J.; Head-Gordon, M.; Trucks, G. W.; Foresman, J. B.; Schlegel, H. B.; Raghavachari, K.; Robb, M.; Binkley, J. S.; Gonzalez, C.; Defrees, D. J.; Fox, D. J.; Whiteside, R. A.; Seeger, R.; Melius, C. F.; Baker, J.; Martin, R. L.; Kahn, L. R.; Stewart, J. J. P.; Topial, S.; Pople, J. A., Gaussian 90, Gaussian Inc., 6823 North Lakewood, Chicago IL 60626
27. Schmidt, M. W.; Baldridge, K. K.; Boatz, J. A.; Jensen, J. H.; Koseki, S.; Gordon, M. S.; Nguyen, K. A.; Windus, T. L.; Albert, S. T. *GAMESS, Quantum Chemistry Program Exchange Bulletin, 10,* **1990**.
28. Amos, R. D.; Rice, J. E. *CADPAC: The Cambridge Analytical Derivatives Package*, Issue 4.0, Cambridge, **1987**.
29. Carbo, R.; Molino, L.; Caloboig, B. *J. Comp. Chem.,* **1992**, *13,*155.
30. Kalamboukis, T. Z. Parallel Computing, **1992**, *18*, 207.
31. Dongarra, J.; Sorensin, D. *SIAM J. Sci. Stat. Computing,* **1987**, *8*, s139.
32. Cuppen, J. J. M. *Numer. Math.,* **1981**, *36*, 177.
33. Ipsen, C. F.; Jessup, E. R. *SIAM J. Sci. Stat. Computing,* **1991**, *12*, 469.
34. Lo, S. S.; Philippe, B.; Sameh, A. *SIAM J. Sci. Stat. Computing,* **1987**, *8*, s155.
35. Demmel, J.; Croz, J. Du.; Hammering, S.; Sorenson, D. , Argonne National Laboratory, MCS-TM-111, **1988**.
36. Szabo, A.; Ostlund, N. S. In *Modern Quantum Chemistry Introduction to Advanced Electronic Structure Theory*; Macmillan: New York, **1982.**
37. MNDO: Dewar, M. J. S.; Thiel, W. *J. Am. Chem. Soc.,* **1977**, *99*, 4899.
38. MNDO/3: Bingham, R. C.; Dewar, M. J. S.; Lo, D. H. *J. Am. Chem. Soc.,* **1975**, *97*, 1294.
39. PM3: Stewart, J. J. P. *J. Comp. Chem.,* **1989**, *10*, 209. Stewart, J. J. P. *J. Am. Chem. Soc.,* **1989**, *10*, 221.
40. Stewart, J. J. P. *J. Comp. Aided Molecular Design,* **1990**, *4*, 45.
41. Jensen, J.; Baldridge, K. K.; Gordon, M. S. *J. Phys. Chem.,* **1992**, *96*, 8340.
42. Bartlet, R. H. University of Texas Center for Numerical Anaylsis, Report CNA-44, Austin, Texas, **1972**.
43. QCPE: Quantum Chemistry Program Exchange, Creative Arts Buiding 181 Indiana University, Bloomington, IN, 47405 USA.
44. CRAY Research, Inc., Unicos Performance Utilities Reference Manual, Publication SR-2040.
45. Press, W. H. In *Numerical Recipes;* Cambridge University Press: New York, **1986**.
46. Gear, C. W. In *Numerical Initial Value Problems in ODE's*; Prentice-Hall, Inc.: New Jersey, **1971**.
47. Gupta, G. K.; Sacks-Davis, R.; Tischer, P. E. *Computer Surveys,* **1985**, *17*, 10.
48. Butcher, J. *Mathematics of Computation,* **1965**, *19*, 408.
49. Burden, R. L.; Faires, J. D.; Reynolds, A. C. In *Numerical Analysis*; Prindle, Weber, and Schmidt: Boston, **1978**.
50. Demmel, J.; Veselic, K. University of Tennessee, CS-89-88,October, **1989**.
51. Yoshitake, Bebbo In *Computers and Chemistry;* **1982**, *6*.
52. Baldridge, K.K. "Parallelization Implementation of Semiempirical Quantum Methods for the Intel Platforms," *Scientific Programming, submitted.*
53. Harrison, R. J. et. al. "Scalable Parallel Algorithms," work done at Pacific Northwest Laboratory, unpublished results.

RECEIVED December 21, 1994

Chapter 9

Parallel Molecular Dynamics Algorithms for Simulation of Molecular Systems

Steve Plimpton and Bruce Hendrickson

Parallel Computational Sciences Department 1421, Mail Stop 1111, Sandia National Laboratories, Albuquerque, NM 87185–1111

Three classes of parallel algorithms for short–range classical molecular dynamics are presented and contrasted and their suitability for simulation of molecular systems is discussed. Performance of the algorithms on the Intel Paragon and Cray T3D in benchmark simulations of Lennard–Jones systems and of a macromolecular system is also highlighted.

Molecular dynamics (MD) is a widely–used tool for simulating liquids and solids at an atomistic level [1]. Molecular systems such as polymers, proteins, and DNA are particularly interesting to study with MD because the conformational shape of the molecules often determines their properties. Such systems are computationally challenging to simulate because (1) in the absence of crystal periodicity large numbers of atoms must often be included in the model, and (2) interesting events such as molecular diffusion or conformational changes typically occur on long timescales relative to the femtosecond–scale timesteps of the MD model.

MD simulations are natural candidates for implementation on parallel computers because the forces on each atom or molecule can be computed independently. MD simulations of molecular systems require computation of two kinds of interactions: *bonded* forces within the topology of the simulated molecules and *non–bonded* van der Waals and Coulombic forces. In this paper we limit our scope to short–range MD models where the non–bonded forces are truncated, so that each atom interacts only with other atoms within a specified cutoff distance. Examples of widely–used commercial and research codes in this category include CHARMM, GROMOS, AMBER, and DISCOVER. While more accurate, MD models with long–range forces are more expensive to compute with, even if hierarchical methods [2] or multipole approximations [13] are used. However, in long–range force models there is a near–field component to the computation which requires a summation of pairwise interactions with near–neighbors. Parallelizing

0097–6156/95/0592–0114$12.00/0

that portion of the computation is essentially the equivalent of the short–range force calculations we will discuss here.

Several techniques have been developed by various researchers for parallelizing short–range MD simulations effectively [11, 15, 20, 24]. The purpose of this paper is to describe the different methods and highlight their respective advantages and disadvantages when applied to molecules, be they small–molecule or macromolecular systems. We begin in the next section with a brief description of the computations that are performed in such MD simulations. The next three sections outline the three basic classes of parallel methods: replicated–data, force–decomposition, and spatial–decomposition approaches. The three methods differ in how they distribute the atom coordinates among processors to perform the necessary computations. Although all of the methods scale optimally with respect to computation, their different data layouts incur different inter–processor communication costs which affect the overall scalability of the methods. In the Results section we briefly describe two benchmark simulations to illustrate the performance and scalability of the parallel methods on two large parallel machines, an Intel Paragon and Cray T3D. The first benchmark is of a Lennard–Jones system with only non–bonded forces; the second is of a solvated myoglobin molecule. Finally, the trade–offs between the three parallel methods are summarized in the conclusion.

Computational Aspects

In MD simulations of molecular systems two kinds of interactions contribute to the total energy of the ensemble of atoms — non–bonded and bonded. These energies are expressed as simple empirical relations [4]; the desired physics or chemistry is simulated by specifying appropriate coefficients. The energy E_{nb} due to non–bonded interactions is typically written as

$$E_{nb} = \sum_i \sum_j \frac{q_i q_j}{r} + \sum_i \sum_j \epsilon_{ij} \left[(\frac{\sigma_{ij}}{r})^{12} - (\frac{\sigma_{ij}}{r})^6 \right] \tag{1}$$

where the first term is Coulombic interactions and the second is van der Waals, r is the distance between atoms i and j, and all subscripted quantities are user–specified constants. In short–range simulations, the summations over i and j are evaluated at each timestep so as to only include atom pairs within a cutoff distance r_c, such that $r < r_c$. The bonded energy E_b for the system in the harmonic approximation can be written as

$$E_b = \sum_{bonds} K_b (r - r_0)^2 + \sum_{angles} K_\theta (\theta - \theta_0)^2 + \sum_{dihedrals} K_{\phi_p} [1 + d_p \cos(n_p \phi)] + \sum_{impropers} K_\phi (\phi - \phi_0)^2 \tag{2}$$

where the first term is 2–body energy, the second is 3–body energy, and the last two are 4–body interactions for torsional dihedral and improper dihedral energies within the topology of the molecules. The distance r and angles θ and ϕ are computed for each interaction as a function of the atomic positions; the subscripted quantities are constants. In contrast to the non–bonded energy, the summations in this equation are explicitly enumerated by the user to setup the simulation, i.e. the connectivities of the molecules are fixed. In the MD simulation, derivatives of Equations 1 and 2 yield force equations for each atom which are integrated over time to generate the motion of the ensemble of atoms.

On a parallel machine with P processors, if a simulation runs P times faster than it does on one processor, it is 100% parallel efficient, or has achieved a perfect speed–up. In molecular simulations both non–bonded and bonded force terms must be spread uniformly across processors to achieve this optimal speed–up. Because atomic densities do not vary greatly in physical systems, the summations in Equation 1 imply that each atom interacts with a small, roughly constant number of neighboring atoms. Similarly, there are a small, fixed number of 2–, 3–, and 4–body interactions in Equation 2 which each atom participates in. Thus, the computational effort in a macromolecular MD model scales linearly with N, the number of atoms in the simulation, and the optimal scaling a parallel method can achieve is as N/P. In any method, whether it scales optimally or not, any exchange of data via inter–processor communication or any imbalance among the processors in computing the terms in Equations 1 or 2 will reduce the parallel efficiency of the method.

Conceptually, the computations in Equation 1 can be represented as a force matrix F where the (ij) element of F is the force due to atom j acting on atom i. The $N \times N$ matrix is sparse due to short–range forces. To take advantage of Newton's 3rd law, we also set $F_{ij} = 0$ when $i > j$ and $i + j$ is even, and likewise set $F_{ij} = 0$ when $i < j$ and $i + j$ is odd. Thus the interaction between a pair of atoms is only computed once. This zeroing of half the matrix elements can also be accomplished by striping F in various ways [23]. Conceptually, F is now colored like a checkerboard with red squares above the diagonal and black squares below the diagonal set to zero. The first two parallel methods we discuss in the next sections assign portions of this F matrix to different processors.

Replicated–Data Method

The most commonly used technique for parallelizing MD simulations of molecular systems is known as the *replicated–data* (RD) method [24]. Numerous parallel algorithms and simulations have been developed based on this approach [5, 8, 9, 16, 17, 18, 22, 25]. Typically, each processor is assigned a subset of atoms and updates their positions and velocities for the duration of the simulation, regardless of where they move in the physical domain.

To explain the method, we first define x and f as vectors of length N which store the position and total force on each atom. Each processor is assigned a sub–block of the force matrix F which consists of N/P rows of the matrix, as shown in Figure 1. If z indexes the processors from 0 to $P - 1$, then processor P_z computes non–bonded forces in the F_z sub–block of rows. It also is assigned the corresponding position and force sub–vectors of length N/P denoted as x_z and f_z. The computation of the non–bonded force F_{ij} requires only the two atom positions x_i and x_j. But to compute all the forces in F_z, processor P_z will need the positions of many atoms owned by other processors. In Figure 1 this is represented by having the horizontal vector x at the top of the figure span all the columns of F. This implies each processor must store a copy of the entire x vector – hence the name replicated–data.

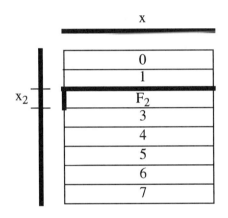

Figure 1: *The division of the force matrix among 8 processors in a replicated–data algorithm. Processor 2 is assigned N/P rows of the matrix and the corresponding x_2 piece of the position vector. In addition, it must know the entire position vector x (shown spanning the columns) to compute the non–bonded forces in F_2.*

The RD algorithm is outlined in Figure 2 with the dominating term in the computation or communication cost of each step listed on the right. We assume at the beginning of the timestep that each processor knows the current positions of all N atoms, i.e. each has an updated copy of the entire x vector. In step (1) of the algorithm, the non–bonded forces in matrix sub–block F_z are computed. This is typically done using neighbor lists to tag the interactions that are likely to be non–zero at a given timestep. In the parallel algorithm each processor would construct lists for its sub–block F_z once every few timesteps. To take advantage of Newton's 3rd law, each processor also stores a copy of the entire force vector f. As each pairwise non–bonded interaction between atoms i and j is computed, the force components are summed twice into f, once in location i and once (negated) in location j, so that F_z is never actually stored as a matrix. Step (1) scales as N/P, the number of non–zero non–bonded interactions computed by each processor.

(1)	**Compute** non–bonded forces in F_z,	
	doubly summing results into local copy of f	$\frac{N}{P}$
(2)	**Compute** $1/P$ fraction of bonded forces,	
	summing results into local copy of f	$\frac{N}{P}$
(3)	**Fold** f across all processors, result is f_z	N
(4)	**Update** atom positions in x_z using f_z	$\frac{N}{P}$
(5)	**Expand** x_z among all processors, result is x	N

Figure 2: *Single timestep of the replicated–data algorithm for processor P_z.*

In step (2) the bonded forces in Equation 2 are computed. This can be done by spreading the loops implicit in the summations of Equation 2 evenly across the processors. Since each processor knows the positions of all atoms, it can compute any of the terms in Equation 2, and sum the resulting forces into its local copy of f. This step also scales as N/P, since there are a small, fixed number of bonded interactions per atom. In step (3), the local force vectors are summed across all processors in such a way that each processor ends up with the total force on each of its N/P atoms. This is the sub–vector f_z. This force summation is a parallel communication operation known as a *fold* [12]. Various algorithms have been developed for performing the operation efficiently on different parallel machines and architectures [3, 12, 26]. The key point is that each processor must essentially receive N/P values from every other processor to sum the total forces on its atoms. The total volume of communication (per processor) is thus $P \times N/P$ and the fold operation thus scales as N.

In step (4), the summed forces are used to update the positions and velocities of each atom. Finally, in step (5) the updated atom positions in x_z are shared among all P processors in preparation for the next timestep. This is essentially the inverse of step (3), and is a communication operation called an *expand* [12]. Since each processor must send its N/P positions to every other processor, this step also scales as N.

The RD algorithm we have outlined divides the MD force computation and integration evenly across the processors; steps (1), (2) and (4) scale optimally as N/P. Load–balance will be good so long as each processor's subset of atoms interacts with roughly the same total number of neighbor atoms. If this does not occur naturally, it can be achieved by randomizing the order of the atoms initially [21] or by adjusting the size of the subsets dynamically as the simulation progresses to tune the load–balance [28]. The chief drawback to the RD algorithm is that it requires global communication in steps (3) and (5); each processor must acquire information held by all the other processors. As indicated above, this communication scales as N, *independent of P*. This means that if the number of processors used in the simulation is doubled, the communication portions of the algorithm do not speed up. Practically speaking this limits the number of processors that can be used effectively.

The chief advantage of the RD algorithm is that of simplicity, particularly for computation of the bonded force terms. The computational steps (1), (2), and (4) can often be implemented by simply modifying the loops and data structures in a serial or vector code to treat N/P atoms instead of N. The fold and expand communication operations (3) and (5) can be treated as black–box routines and inserted at the proper locations in the code. Few other changes are typically necessary to parallelize an existing code.

Force–Decomposition Method

A parallel algorithm that retains many of the advantages of the replicated–data approach, while reducing its communication costs, can be formulated by partitioning the force matrix F by sub–blocks rather than rows, as illustrated in Figure 3. We call this a *force–decomposition* (FD) method [20]. Use of the method in the macromolecular MD code ParBond is described in [21]; a modified FD approach has also been implemented in a parallel version of CHARMM [6].

The block–decomposition in Figure 3 is actually of a permuted force matrix F' which is formed by rearranging the columns of the original checkerboarded F in a particular way. As before, we let z index the processors from 0 to $P-1$; processor P_z owns and will update the N/P atoms stored in the sub–vector x_z. If we order the x_z pieces in *row–order* (across the rows of the matrix), they form the usual position vector x which is shown as a vertical bar at the left of the figure. Were we to have x span the columns as in Figure 1, we would form the force matrix as before. Instead, we span the columns with a permuted position vector x', shown as a horizontal bar at the top of Figure 3, in which the x_z pieces are stored in *column–order* (down the columns of the matrix). Thus, in the 16–processor example shown in the figure, x stores each processor's piece in the usual order $(0, 1, 2, 3, 4, ..., 14, 15)$ while x' stores them as $(0, 4, 8, 12, 1, 5, 9, 13, 2, 6, 10, 14, 3, 7, 11, 15)$. Now the (ij) element of F' is the force on atom i in vector x due to atom j in permuted vector x'.

The F'_z sub–block owned by each processor P_z is of size $(N/\sqrt{P}) \times (N/\sqrt{P})$. As indicated in the figure, to compute the non–bonded forces in F'_z, processor P_z must know one $N/\sqrt{P}$–length piece of each of the x and x' vectors, which we denote as x_α and x'_β. As these elements are computed they will be accumulated into corresponding force sub–vectors f_α and f'_β. The Greek subscripts α and β each run from 0 to $\sqrt{P}-1$ and reference the row and column position occupied by processor P_z. Thus for processor 6 in the figure, x_α consists of the x sub–vectors $(4, 5, 6, 7)$ and x'_β consists of the x' sub–vectors $(2, 6, 10, 14)$.

The FD algorithm is outlined in Figure 4. As before, each processor has updated copies of the needed atom positions x_α and x'_β at the beginning of the

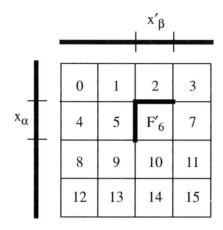

Figure 3: *The division of the permuted force matrix F' among 16 processors in the force–decomposition algorithm. Processor P_6 is assigned a sub–block F'_6 of size $N/\sqrt{P}$ by $N/\sqrt{P}$. To compute the non–bonded forces in F'_6 it must know the corresponding $N/\sqrt{P}$-length pieces x_α and x'_β of the position vector x and permuted position vector x'.*

timestep. In step (1), the non–bonded forces in matrix sub–block F'_z are computed. As before, neighbor lists can be used to tag the $O(N/P)$ non–zero interactions in F'_z. As each force is computed, the result is summed into the appropriate locations of both f_α and f'_β to account for Newton's 3rd law. In step (2) each processor computes an N/P fraction of the bonded interactions. Since each processor knows only a subset of atom positions, this must be done differently than in the RD algorithm. For each set of 2, 3, or 4 atoms corresponding to a bonded interaction term in Equation 2, we must guarantee that some processor knows all the needed atom positions. This can be accomplished be ordering the atoms in the x vector appropriately as a pre–processing step before the MD simulation is begun. A heuristic method for doing this is described in reference [21].

In step (3), the force on each processor's atoms is acquired. The total force on atom i is the sum of elements in row i of the force matrix minus the sum of elements in column i', where i' is the permuted position of column i. Step (3a) performs a fold within each row of processors to sum the first of these contributions. Although the fold algorithm used is the same as in the previous section, there is a key difference. In this case the vector f_α being folded is only of length $N/\sqrt{P}$ and only the $\sqrt{P}$ processors in one row are participating in the fold. Thus this operation scales as $N/\sqrt{P}$ instead of N as in the RD algorithm. Similarly, in step (3b), a fold is done within each column of F'. The two contributions to the total force are joined in step (3c).

(1) **Compute** non–bonded forces in F'_z, storing results in f_α and f'_β	$\frac{N}{P}$
(2) **Compute** $1/P$ fraction of bonded forces, storing results in f_α and f'_β	$\frac{N}{P}$
(3a) **Fold** f_α within row α, result is f_z	$\frac{N}{\sqrt{P}}$
(3b) **Fold** f'_β within column β, result is f'_z	$\frac{N}{\sqrt{P}}$
(3c) **Subtract** f'_z from f_z, result is total f_z	$\frac{N}{P}$
(4) **Update** atom positions in x_z using f_z	$\frac{N}{P}$
(5a) **Expand** x_z within row α, result is x_α	$\frac{N}{\sqrt{P}}$
(5b) **Expand** x_z within column β, result is x'_β	$\frac{N}{\sqrt{P}}$

Figure 4: *Single timestep of the force–decomposition algorithm for processor P_z.*

In step (4), f_z is used to update the N/P atom positions in x_z. Steps (5a–5b) share these updated positions with all the processors that will need them for the next timestep. These are the processors which share a row or column with P_z. First, in (5a), the processors in row α perform an expand of their x_z sub–vectors so that each acquires the entire x_α. As with the fold, this operation scales as the $N/\sqrt{P}$ length of x_α instead of as N as it did in the RD algorithm. Similarly, in step (5b), the processors in each column β perform an expand of their x_z. As a result they all acquire x'_β and are ready to begin the next timestep.

As with the RD method, the FD method we have outlined divides the MD computations evenly among the processors. Step (1) will be load–balanced if all the matrix sub–blocks F'_z are uniformly sparse. As with the RD method, a randomized initial ordering of atoms produces the desired effect. The key enhancement offered by the FD method is that the communication operations in steps (3) and (5) now scale as $N/\sqrt{P}$ rather than as N as was the case with the RD algorithm. When run on large numbers of processors this can significantly reduce the time spent in communication. Likewise, memory costs for position and force vectors are reduced by the same $\sqrt{P}$ factor. Finally, though more steps are needed, the FD approach retains the overall simplicity and structure of the RD method; it can be implemented using the same expand and fold communication routines.

Spatial–Decomposition Method

The final parallel method we describe exploits the locality of the short–range forces by assigning to each of the P processors a small region of the simulation domain. As illustrated in Figure 5 this is a *geometric–* or *spatial–decomposition* (SD) of the workload. For reasons that will be outlined below, there have been fewer implementations of short–range macromolecular MD simulations using this method [7, 10, 27] than with RD approaches.

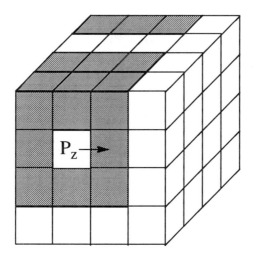

Figure 5: *The division of the 3–D periodic simulation domain among 64 processors in a spatial–decomposition algorithm. With this force cutoff distance (arrow), processor P_z only need communicate with the 26 nearest–neighbor processors owning the shaded boxes.*

The SD algorithm is outlined in Figure 6. Processor z owns the box labeled D_z and will update the positions x_z of the atoms in its box. To compute forces on its atoms a processor will need to know not only x_z but also positions y_z of atoms owned by processors whose boxes are within a cutoff distance r_c of its box. As it computes the forces f_z on its atoms, it will also compute components of forces g_z on the nearby atoms (Newton's 3rd law).

We again assume that current x_z and y_z positions are known by each processor at the beginning of the timestep. With these definitions, steps (1) and (2) of the algorithm are the computation of non–bonded and bonded forces for interactions involving the processor's atoms. These steps scale as the number of atoms N/P in each processor's box. In step (3) the g_z forces computed on neighboring atoms are communicated to processors owning neighboring boxes. The received forces are summed with the previously computed f_z to create the total force on a processor's atoms. The scaling of this step depends on the length of the force cutoff relative to the box size. We list it as Δ and discuss it further below. Step (4) updates the positions of the processor's atoms. In step (5) these positions are communicated to processors owning neighboring boxes so that all processors can update their y_z list of nearby atoms. Finally in step (6), periodically (usually when neighbor lists are created), atoms which have left a processor's box must be moved to the appropriate new processor.

The above description ignores many details of an effective SD algorithm [20],

(1) **Compute** non–bonded forces in D_z, summing results into f_z and g_z	$\frac{N}{P}$
(2) **Compute** bonded forces in D_z, summing results into f_z and g_z	$\frac{N}{P}$
(3) **Share** g_z with neighboring processors, summing received forces into my f_z	Δ
(4) **Update** atom positions in x_z using f_z	$\frac{N}{P}$
(5) **Share** x_z with neighboring processors, using received positions to update y_z	Δ
(6) **Move** atoms to new processors as necessary	Δ

Figure 6: *Single timestep of the spatial–decomposition algorithm for processor P_z.*

but it is clear that the computational scaling of steps (1), (2), and (4) is again the optimal N/P. The scaling of the communication steps (3), (5), and (6) is more complex. In the limit of large N/P ratios, Δ scales as the surface–to–volume ratio $(N/P)^{(2/3)}$ of each processor's box. If each processor's box is roughly equal in size to the force cutoff distance, then Δ scales as N/P and each processor receives N/P atom positions from each of its neighboring 26 processors (in 3–D), as in Figure 5. In practice, however, there can be several obstacles to minimizing Δ and achieving high parallel efficiencies for a SD method in MD simulations of molecular systems.

(A) Molecular systems are often simulated in a vacuum or with surrounding solvent that does not uniformly fill a 3–D box. In this case it is non–trivial to divide the simulation domain so that every processor's box has an equal number of atoms in it and yet keep the inter–processor communication simple. Load–imbalance is the result.

(B) Because of the $1/r$ dependence of Coulombic energies in Equation 1, long cutoffs are often used in simulations of organic materials. Thus a processor's box may be much smaller than the cutoff. The result is considerable extra communication in steps (3) and (5) to acquire needed atom positions and forces, i.e. Δ no longer scales as N/P, but as the cube of the cutoff distance r_c.

(C) As atoms move to new processors in step (6), molecular connectivity information must be exchanged and updated between processors. The extra coding to manipulate the appropriate data structures and optimize the communication performance of the data exchange subtracts from the parallel efficiency of the algorithm.

In general, SD methods are more difficult to integrate into large, existing codes than are RD or even FD methods. This fact, coupled with the potential for other parallel inefficiencies just outlined (A-C), has made SD implementations less common than RD for macromolecular MD codes. However, in terms of their optimal communication scaling they are clearly the method of choice for very large simulations. Additionally, for MD codes that include long–range force interactions via multipole methods [13], SD methods are a natural choice for performing the near–field pairwise computations. This is because the far–field multipole contributions to the forces are computed on a hierarchy of spatial grids that correspond to coarsened versions of the fine grid pictured in Figure 5.

Results

In this section we highlight the performance of the three parallel MD algorithms just described on two large parallel machines, an 1840–processor Intel Paragon at Sandia running the SUNMOS operating system [19] and a 512–processor Cray T3D at Cray Research. The same F77 codes were run on both machines; only a few lines of code are machine–specific calls to native send and receive message–passing routines.

The first set of results are for a benchmark simulation of Lennard–Jonesium [20]; just the second non–bonded term in Equation 1 is included in the force model. A 3–D periodic box of atomic liquid is simulated with a standard force cutoff of 2.5σ encompassing an average of 55 neighbors/atom. The CPU time per MD timestep is shown in Figure 7 for runs of various sized systems on single processors of the Cray Y–MP and C90 and on 1024 processors of the Intel Paragon. The code run on the Y–MP and C90 is a slightly modified version of the algorithm of Grest et al. [14] which vectorizes completely and has produced the fastest timings to date for this benchmark on conventional vector supercomputers [14, 20]. The Paragon timings are for codes which implement the three parallel algorithms discussed in the previous sections: replicated–data (RD), force–decomposition (FD), and spatial–decomposition (SD); more details are given in reference [20]. The three timing curves for 512 processors of the Cray T3D are virtually identical to these (to within a few percent), meaning the T3D's computation and communication rates for these codes are twice as fast as the Paragon on a per–processor basis.

The data in the figure show that, as expected, all of the algorithms scale linearly with N in the large N limit. The timings for FD are faster than RD for all sizes, due to FD's reduced communication cost. For small problems, FD is the fastest of the three parallel algorithms; for larger sizes SD quickly becomes the fastest method. For large N the difference in timings between the three algorithms is due to their relative communication costs; all of them are essentially equivalent with respect to parallelizing the computational portions of the timestep calculation.

It is worth noting that this is a benchmark problem for which the SD approach is ideally suited. The simulated atoms uniformly fill a 3–D box which can be easily partitioned equally among the processors. More irregular problems would lead to load–imbalance which would reduce the parallel efficiency of the SD method, but not the FD and RD methods. Also, the crossover size at which SD becomes faster than FD is a function of P and of several features of the benchmark, in

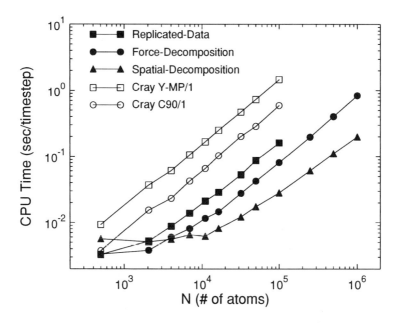

Figure 7: *CPU timings (seconds/timestep) for the three parallel algorithms on 1024 processors of the Intel Paragon for different problem sizes. Single–processor Cray Y–MP and C90 timings are also given for comparison.*

particular the cutoff distance. Consider the case where the cutoff is increased to 5.0σ to encompass 440 neighbors/atom. This is more typical of the cutoffs used in charged systems to include more of the longer–range Coulombic interactions. There is now 8 times as much computation per atom to be performed in all of the parallel methods. In the SD method there is also 8 times as much communication to acquire atoms within the cutoff distance, so the ratio of communication to computation is unchanged. By contrast, in the RD and FD methods, the amount of communication is *independent* of the cutoff distance, so the ratio of communication to computation is reduced and the parallel efficiency of the methods goes up. The net effect is to shift the crossover size where SD becomes faster to larger N. In practice this can be many tens of thousands of atoms [20].

The timing data in Figure 7 also indicate what is feasible on current–generation parallel supercomputers for short–range MD simulations of Lennard–Jones systems. On the million–atom simulation the 1024–processor Paragon is running at 0.199 seconds/timestep, about 30 times faster than a C90 processor (extrapolated). Similarly the 512–processor T3D runs at 0.205 seconds/timestep. If all 1840 nodes of Sandia's Paragon are used, if the dual–processor mode is enabled where a second i860 processor on each node normally used for communication is used for computation, and if the Lennard–Jones force computation kernel is written in assembler rather than Fortran, these timing numbers can be improved by about a factor of 4–5 for large N [20]. The million–atom simulation then runs at 0.045 seconds/timestep (80,000 timesteps/hour) and 100,000,000 atoms can be simulated in 3.53 seconds/timestep, about 165 times faster than a C90 processor.

Timing results for a macromolecular simulation of myoglobin using the force model of Equations 1 and 2 are shown in Figure 8. This is a prototypical protein benchmark proposed by Brooks et al. [5] who have done extensive testing of a variety of machines with CHARMM for this problem. A 2534–atom myoglobin molecule (with an adsorbed CO) is surrounded by a shell of solvent water molecules for a total of 14,026 atoms. The resulting ensemble is roughly spherical in shape. The benchmark is a 1000–timestep simulation performed at a temperature of 300° K with a non–bonded force cutoff of 12.0 Å. Neighbor lists are created every 25 timesteps with a 14.0 Å cutoff.

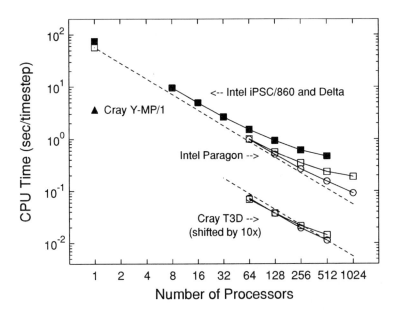

Figure 8: *CPU timings (seconds/timestep) on different numbers of processors for a 14,026–atom myoglobin benchmark. The squares are timings for replicated–data implementations; circles are for force–decomposition. Timings for the filled symbols are from reference [5].*

All of the filled symbols in the figure are timings due to Brooks et al. [5].

The single processor Cray Y–MP timing of 3.64 secs/timestep is for a version of CHARMM they have optimized for vector processing. They have also developed a parallel version of CHARMM [5] using a RD algorithm similar to out replicated–data method. Timings with that version on an Intel iPSC/860 and the Intel Delta at CalTech are shown in the figure as filled squares.

We have implemented both RD and FD algorithms in a parallel MD code for molecular systems we have written called ParBond. It is similar in concept (though not in scope) to the widely–used commercial and academic macromolecular codes CHARMM, AMBER, GROMOS, and DISCOVER. In fact, ParBond was designed to be CHARMM–compatible in the sense that it uses the same force equations as CHARMM [4]. Since the RD and FD methods both use the same communication primitives, ParBond simply has a switch that partitions the force matrix either by rows or sub–blocks as in Figures 1 and 3.

Timings for ParBond on the Intel Paragon using the RD and FD described earlier are shown by the upper set of open squares and circles respectively in the figure. Taking into account that the i860XP floating point processors in the Paragon are about 30% faster than the i860XR chips in the iPSC/860 and Delta and that inter–processor communication is significantly faster on the Paragon, the two sets of RD timings (filled and open squares) are similar. Both curves show a marked roll–off in parallel efficiency above 64–128 processors due to the poor scaling of the expand and fold operations. This is typical of the results reported in references [8, 9, 16, 17, 18, 22, 25] for RD implementations of other macromolecular codes such as CHARMM, AMBER, and GROMOS on a variety of parallel machines. Parallel efficiencies as low as 10–15% on a few dozens to hundreds of processors are reported and in some cases the overall speed–up is even reduced as more processors are added due to communication overheads. The implementation of Sato et al. [22] is a notable exception which achieves parallel efficiencies of 32 and 44% on 512 processors for their two benchmark calculations.

By contrast the FD algorithm timings in ParBond (open circles) for the Paragon fall off less rapidly as processors are added; it is running 1.3 times faster than its RD counterpart on 256 processors (0.265 secs/timestep vs. 0.347) and 2.1 times faster on 1024–processors (0.0913 secs/timestep vs. 0.189). In other macromolecular simulations it has performed up to 3.3 times faster than the RD algorithm on 1024 Paragon processors [21]. The 1024–processor Paragon timing for FD in Figure 8 is about 40 times faster than the single Y–MP processor timing. The long dotted line in the figure represents perfect speed–up or 100% parallel efficiency for the ParBond code on the Paragon extrapolated from an estimated one–processor timing. The FD algorithm still has a relatively high parallel efficiency of 61% on 1024 processors, as compared to 30% for the RD timing.

Cray T3D timings on 64–512 processors are also shown in the figure for Par-Bond using the RD and FD algorithms (lower set of open squares and circles). They are shifted downward be a factor of 10 so as to not overlay the Paragon data. A short dotted reference line is provided by shifting the Paragon perfect speed–up line down by a factor of 10 as well. The FD timings on the T3D are about 25%

faster than their Paragon counterparts on the same number of processors. This is less than the 2x factor on the Lennard–Jones benchmark for two reasons. First, the ratio of single–processor computation rates between the T3D and Paragon is not as high for ParBond, indicating more optimization work needs to be done on the T3D version of the code. Second, the computation–to–communication ratio is higher in this benchmark because of the longer cutoffs and more complicated force equations. Thus there is relatively less time spent in communication, and the T3D's higher effective communication bandwidth (due to its 3–D torus topology vs. a 2–D mesh for the Paragon) is less of a factor. These are also the reasons the FD algorithm is less of a win relative to a RD approach on the T3D; the advantage of FD is its communication scaling and the RD implementation on the T3D is only spending a small fraction of its time in inter–processor communication.

We are not aware of any spatial–decomposition (SD) implementations of this myoglobin benchmark to compare with the RD and FD results presented here. Because the atoms fill a spherical volume instead of a box and because the cutoff distance is relatively long (950 neighbors/atom), we would expect an SD approach on hundreds or thousands of processors to have difficulty matching the 61% parallel efficiency of the FD algorithm on 1024 Paragon processors for this problem. However, in principle, for larger (or more uniform) molecular systems even higher efficiencies should be possible with SD methods. We briefly describe three notable efforts in this area.

Esselink and Hilbers have developed their SD model [10] for a 400–processor T800 Transputer machine. They partition uniform domains in 2–D columns with the 3rd dimension owned wholly on processor and have achieved parallel efficiencies on regular problems of as high as 50%. Clark et al. have implemented a more robust 3–D SD strategy in their recently developed EulerGROMOS code [7]. By recursively halving the global domain across subsets of processors, each processor ends up with a rectangular–shaped sub–domain of variable size which may not align with its neighbors. This allows irregular–shaped global domains to be partitioned across processors in a load–balanced fashion at the cost of extra communication overhead. They report a parallel efficiency of roughly 10% on 512 processors of the Intel Delta at CalTech for a 10914–atom benchmark computation of solvated myoglobin with a 10.0 Å cutoff in a uniformly filled 3–D box. Finally, Windemuth, has also implemented a novel solution to the load–balancing problem for irregular–shaped domains in his SD code PMD [27]. He defines one Voronoi point per processor scattered throughout the simulation domain. The domain is then tesselated so that each processor ends up owning the physical region of volume closest to its Voronoi point. By adjusting the position of the Voronoi points as the simulation progresses and re–tesselating, the simulation can keep the volume (work) per processor roughly constant and thus insure continued load–balance even for non–uniform atom densities.

Conclusions

We have discussed three methods suitable for parallelizing MD simulations of molecular systems. Their basic characteristics are summarized in Figure 9. The scalability of the computation and communication portions of the algorithms are listed in the first two columns as a function of number of atoms N and number of processors P. To first order all the methods parallelize the MD computation optimally as N/P. (Strictly speaking, this ignores some costs in neighbor list construction which are typically small contributions to the total computational cost [20]). The chief difference in the methods is in communication cost where the SD method is a clear winner in the large N limit. Memory costs are listed in the 3rd column. In practice, the $O(N)$ cost of RD methods can limit the size of problems that can be simulated [5, 21], while on current parallel machines the FD and SD methods are more limited by compute power than by memory.

Method	Computation	Communi-cation	Memory	Ease of Coding	Load Balance
RD	$\dfrac{N}{P}$	N	N	simple	geometry-insensitive
FD	$\dfrac{N}{P}$	$\dfrac{N}{\sqrt{P}}$	$\dfrac{N}{\sqrt{P}}$	moderate	geometry-insensitive
SD	$\dfrac{N}{P}$	$\left(\dfrac{N}{P}\right)^{2/3}$	$\dfrac{N}{P}$	complex	geometry-sensitive

Figure 9: *Comparative properties of three parallel methods for short–range molecular dynamics simulations: replicated–data (RD), force–decomposition (FD), and spatial–decomposition (SD). The scalability of the algorithm's computation, communication, and memory requirements when simulating N atoms on P processors is listed. The relative ease of implementation and load–balancing characteristics of the three methods are also shown.*

In the fourth column the relative ease of coding or implementing the three methods for molecular simulations is listed; RD is the most straightforward, FD requires more work to handle bonded interactions correctly, and SD is the most complex of the three in terms of data structures and communication of molecular connectivity as molecules move from processor to processor. Finally, the load–balancing properties of the methods are listed in the last column. Both the RD and FD methods are geometry–insensitive, meaning the processor's workload does not change as atoms move within the physical domain. In other words, simulations of irregular–shaped domains are no more difficult to load–balance than regular domains. By contrast, SD methods are sensitive to the spatial location of the molecules. Load–imbalance can result if particle densities are non–uniform.

Acknowledgments

This work was partially supported by the Applied Mathematical Sciences program, U.S. Department of Energy, Office of Energy Research, and was performed at Sandia National Laboratories, operated for the DOE under contract No. DE–AC04–76DP00789. The Cray T3D and C90 simulations were performed on machines at Cray Research. We thank Barry Bolding of Cray Research and John Mertz (now at the Minnesota Supercomputer Center) for assisting in that effort. We also thank Gary Grest of Exxon Research for providing us a copy of his vectorized Lennard–Jones code for performing the Cray benchmark calculations discussed in the same section.

References

[1] M. P. Allen and D. J. Tildesley. *Computer Simulation of Liquids.* Clarendon Press, Oxford, 1987.

[2] J. E. Barnes and P. Hut. A hierarchical $O(N \log N)$ force–calculation algorithm. *Nature*, 324:446–449, 1986.

[3] M. Barnett, L. Shuler, R. van de Geijn, S. Gupta, D. Payne, and J. Watts. Interprocessor collective communication library (Intercom). In *Proc. Scalable High Performance Computing Conference–94*, pages 357–364. IEEE Computer Society Press, 1994.

[4] B. R. Brooks, R. E. Bruccoleri, B. D. Olafson, D. J. States, S. Swaminathan, and M. Karplus. CHARMM: A program for macromolecular energy, minimization, and dynamics calculations. *J. Comp. Chem.*, 4:187–217, 1983.

[5] B. R. Brooks and M. Hodošček. Parallelization of CHARMM for MIMD machines. *Chemical Design Automation News*, 7:16–22, 1992.

[6] Brooks, B. R. at National Institutes of Health, personal communication, 1994.

[7] T. W. Clark, R. V. Hanxleden, J. A. McCammon, and L. R. Scott. Parallelizing molecular dynamics using spatial decomposition. In *Proc. Scalable High Performance Computing Conference–94*, pages 95–102. IEEE Computer Society Press, 1994.

[8] T. W. Clark, J. A. McCammon, and L. R. Scott. Parallel molecular dynamics. In *Proc. 5th SIAM Conference on Parallel Processing for Scientific Computing*, pages 338–344. SIAM, 1992.

[9] S. E. DeBolt and P. A Kollman. AMBERCUBE MD, Parallelization of AMBER's molecular dynamics module for distributed–memory hypercube computers. *J. Comp. Chem.*, 14:312–329, 1993.

[10] K. Esselink and P. A. J. Hilbers. Efficient parallel implementation of molecular dynamics on a toroidal network: II. Multi–particle potentials. *J. Comp. Phys.*, 106:108–114, 1993.

[11] D. Fincham. Parallel computers and molecular simulation. *Molec. Sim.*, 1:1–45, 1987.

[12] G. C. Fox, M. A. Johnson, G. A. Lyzenga, S. W. Otto, J. K. Salmon, and D. W. Walker. *Solving Problems on Concurrent Processors: Volume 1*. Prentice Hall, Englewood Cliffs, NJ, 1988.

[13] L. Greengard and V. Rokhlin. A fast algorithm for particle simulations. *J. Comp. Phys.*, 73:325–348, 1987.

[14] G. S. Grest, B. Dünweg, and K. Kremer. Vectorized link cell Fortran code for molecular dynamics simulations for a large number of particles. *Comp. Phys. Comm.*, 55:269–285, 1989.

[15] S. Gupta. Computing aspects of molecular dynamics simulations. *Comp. Phys. Comm.*, 70:243–270, 1992.

[16] H. Heller, H. Grubmuller, and K. Schulten. Molecular dynamics simulation on a parallel computer. *Molec. Sim.*, 5:133–165, 1990.

[17] J. F. Janak and P. C. Pattnaik. Protein calculations on parallel processors: II. Parallel algorithm for forces and molecular dynamics. *J. Comp. Chem.*, 13:1098–1102, 1992.

[18] S. L. Lin, J. Mellor-Crummey, B. M. Pettit, and G. N. Phillips Jr. Molecular dynamics on a distributed–memory multiprocessor. *J. Comp. Chem.*, 13:1022–1035, 1992.

[19] A. B. Maccabe, K. S. McCurley, R. Riesen, and S. R. Wheat. SUNMOS for the Intel Paragon: A brief user's guide. In *Proceedings of the Intel Supercomputer User's Group. 1994 Annual North America Users' Conference.*, 1994.

[20] S. J. Plimpton. Fast parallel algorithms for short–range molecular dynamics. *J. Comp. Phys.*, 1994. To appear.

[21] S. J. Plimpton and B. A. Hendrickson. A new parallel method for molecular dynamics simulation of macromolecular systems. Technical Report SAND94–1862, Sandia National Laboratories, Albuquerque, NM, 1994. Submitted for publication.

[22] H. Sato, Y. Tanaka, H. Iwama, S. Kawakika, M. Saito, K. Morikami, T. Yao, and S. Tsutsumi. Parallelization of AMBER molecular dynamics program for the AP1000 highly parallel computer. In *Proc. Scalable High Performance Computing Conference–92*, pages 113–120. IEEE Computer Society Press, 1992.

[23] H. Schreiber, O. Steinhauser, and P. Schuster. Parallel molecular dynamics of biomolecules. *Parallel Computing*, 18:557–573, 1992.

[24] W. Smith. Molecular dynamics on hypercube parallel computers. *Comp. Phys. Comm.*, 62:229–248, 1991.

[25] W. Smith and T. R. Forester. Parallel macromolecular simulations and the replicated data strategy: I. The computation of atomic forces. *Comp. Phys. Comm.*, 79:52–62, 1994.

[26] R. van de Geijn. Efficient global combine operations. In *Proc. 6th Distributed Memory Computing Conference*, pages 291–294. IEEE Computer Society Press, 1991.

[27] A. Windemuth. *Advanced algorithms for molecular dynamics simulation: The program PMD.* published by the American Chemical Society, 1994. In this volume.

[28] W. S. Young and C. L. Brooks, III. Dynamic load balancing algorithms for replicated data molecular dynamics, 1994. Submitted for publication.

RECEIVED November 15, 1994

Chapter 10

Portable Molecular Dynamics Software for Parallel Computing

Timothy G. Mattson[1] and Ganesan Ravishanker[2]

[1]Intel Corporation, Supercomputer Systems Division, Mail Stop C06-09, 14924 Northwest Greenbrier Parkway, Beaverton, OR 97009
[2]Department of Chemistry, Hall-Atwater Labs, Wesleyan University, Middletown, CT 06457

In this paper, we describe a parallel version of Wesdyn; a molecular dynamics program based on the GROMOS force field. Our goal was to approach the parallelization as software engineers and focus on portability, maintainability, and ease of coding. These criteria were met with an SPMD, loop-splitting algorithm that used a simple owner-computes-filter to assign loop iterations to the nodes of the parallel computer. The program was implemented with TCGMSG and Fortran-Linda and was portable among MIMD parallel computers. We verified the portability by running on several different MIMD computers, but only report workstation cluster results in this chapter.

Molecular dynamics (MD) simulations are extremely compute intensive and require supercomputer performance to provide answers in a reasonable amount of time. Given the high cost of supercomputers, there is a great deal of interest among the users of MD software to utilize the most cost effective supercomputers – those based on parallel and distributed architectures.

The algorithms needed to utilize parallel computing are well understood [5], so one might expect parallel computers to play a dominant role in MD. In practice, however, very few MD users utilize parallel systems. The reason for this is simple – while the hardware for parallel computing is readily available, the application software isn't. This state of affairs is due to the *dusty deck* problem: users of MD simulations depend on established programs that are not officially supported for execution on parallel systems. The problems of dealing with these codes is further complicated because they were developed long before the advent of parallel computers and are poorly structured for these systems.

0097–6156/95/0592–0133$12.00/0

One solution to the *dusty deck* problem is to replace the old programs with new software designed to take advantage of parallel computers. MD programs, however, have been painstakingly developed over the years and have attracted a dedicated following. Those using these codes are unlikely to replace their tried-and-true MD programs with modern, unfamiliar codes. Hence, there is no way around the painful reality: to move MD simulations onto parallel computers, these old dusty deck programs must be ported to parallel systems.

In principle, one should have to parallelize only one of the popular MD programs and reuse the computational kernels among the other codes to yield a full set of parallel MD programs. Unfortunately, this is not practical. MD programs are tightly inter-twined with their force fields. These force fields embody the science of the MD codes and have been designed for different types of problems. The result is that the MD user community needs all the various MD programs so each of these must be parallelized separately with little code sharing.

Given the need to port so many established MD programs to multiple parallel computers, the major outstanding problems in parallel MD pertain to software engineering – not algorithm design. Therefore, we decided to explore parallelization of MD programs from a software engineer's perspective. Our first test case is WESDYN [3] – a molecular dynamics program based on the GROMOS [2] force field. We set the following guidelines for the parallelization project:

- Design the parallel algorithm so any changes to the sequential code will be simple and well isolated.

- Implement the changes so the parallel and sequential programs can be maintained within the same source code.

- Support portable execution of the parallel code among different MIMD parallel systems.

While important, performance of the parallel program was a secondary concern relative to the issues of quality software engineering. In addition, the primary target was modest numbers of clustered workstations, though we wanted to write portable code that could easily move to moderate sized MIMD parallel computers as well.

We believe that focusing on moderate parallelization is appropriate relative to the needs of the broadest category of users. While the National Supercomputer Centers provide some access to supercomputers, the available time on these machines is scarce. By producing a well engineered code that runs on modestly sized parallel systems, we provide a program that benefits the majority of users.

The paper begins by describing the sequential WESDYN program. This is followed by a discussion of the parallel version of the code using Fortran-Linda [11] and TCGMSG [6]. While we verified portability by running on several parallel systems, we only present results for workstation clusters in this chapter. This decision was made since the results for the MIMD computers are preliminary and require additional analysis prior to publication.

The Sequential WESDYN program

WESDYN [3] is a molecular dynamics program which uses the GROMOS [2] force field. The program has been heavily used at a number of sites, particularly for the simulations of DNA, proteins and protein-DNA complexes. It is particularly well known for is its optimization for vector supercomputers from Cray Research.

Before describing the program in detail, consider a high level view of a molecular dynamics simulation. Molecular dynamics programs simulate atomic motions within a molecular system by using a simplified representation of the molecular forces. In the course of the computation, the molecular system evolves through many successive time steps where for each time step:

- Compute bonded energies and forces.

- Compute non-bonded energies and forces.

- Integrate classical equations of motion to propagate to the next time step.

While the non-bonded energies include terms that require all pairs of atoms to interact, in most MD programs, only those atoms within a preset cutoff distance are included. This list of neighbors for the non-bonded calculation is computed every 10 to 50 steps. In addition, it is sometimes important to constrain the system to known physical limits using the shake algorithm [13] and/or velocity scaling.

Finally, most intesting molecular processes do not take place in a vacuum so a large number of solvent molecules must be included within the simulated system. To prevent artifacts in the simulation due to the size of the simulation box, appropriate boundary conditions must be applied to the system.

In WESDYN, Hexagonal Prizm Boundary (HPB) conditions [4] are used. The HPB algorithm applies periodic boundary conditions across the faces of hexagonal prizm unit cells. The first set of neighboring cells are along each of the six sides of the prizm. This 7 cell ensemble is then stacked on top and on bottom to give a total of 20 neighbors for any cell.

In the figure 1, we provide pseudo code for WESDYN. The program begins with the input of user data. Based on this input, the program carries out a number of energy minimization steps to reduce strain within the initial structure. This energy minimization uses the same force field as is used in the dynamics simulation itself – even to the point of calling many of the same energy routines.

Once a low energy initial structure has been found, the molecular dynamics time-stepping loop is entered. For the first step and at fixed intervals thereafter,

```
program WESDYN
      PROCESS_USER_COMMANDS
      READ_MOLECULAR_TOPOLOGY_FILE
      READ_CARTESIAN_COORDINATES
      READ_INITIAL_VELOCITIES
      MINIMIZE_ENERGY_OF_INITIAL_STRUCTURE
      do i = 1,number_of_steps
         if (TIME_TO_UPDATE_NON_BONDED_NEIGHBOR_LIST) then
                  EVALUATE_NON_BONDED_NEIGHBOR_LIST
         endif
         CALCULATE_INTERNAL_COORDINATE_ENERGIES_AND_FORCES
         CALCULATE_GEOMETRIC_CENTERS_OF_CHARGE_GROUPS
         CALL_NBSTST   ! evaluate solute-solute energies and forces
         CALL_NBSTSV   ! evaluate solute-solvent energies and forces
         CALL_NBSVSV   ! evaluate solvent-solvent energies and forces
         INTEGRATE_EQUATIONS_OF_MOTION
         if (SHAKE_REQUESTED) then
            APPLY_SHAKE
         endif
         if (TEMPERATURE_OFF_LIMITS) then
            RESCALE_VELOCITIES
         endif
      end do
end WESDYN
```

Figure 1: Pseudo-code description of the original sequential version of WESDYN

```
subroutine NBSTST
     do i = 1,number_of_charge_groups
        number_of_interacting_groups =
               COLLECT_ALL_GROUPS_WITHIN_CUTOFF
        do j = 1,number_of_interacting_groups
           SWITCH =
             CALCULATE_SWITCHING_FUNCTION(group(i),group(j))
           do k = 1,number_of_atoms(group(i))
              do l = 1,number_of_atoms(group(j))
                 E = E + SWITCH *
                        ENERGY(atom(k,group(i)),atom(l,group(j)))
                 F_PAIR = FORCE(atom(k,group(i)),atom(l,group(j)))
                 F(atom(k,group(i))) = F(atom(k,group(i)))
                             + SWITCH * F_PAIR
                 F(atom(l,group(j))) = F(atom(l,group(j)))
                             - SWITCH * F_PAIR
              end do
           end do
        end do
     end do
end NBSTST
```

Figure 2: Pseudo-code description of NBSTST.

a list of groups within a certain cutoff distance is computed. This list plays a key role in the later computation of non-bonded forces and energies. For every time step, internal energies and forces are computed. This is followed by the computational core of the program – the computation of non-bonded forces and energies. This computation is split between three routines:

- NBSTST: Non-bonded routine for solute-solute interactions.

- NBSTSV: Non-bonded routine for solute-solvent interactions.

- NBSVSV: Non-bonded routine for solvent-solvent interactions.

These non-bonded energy routines loop over all charge groups within the molecule. All groups within the cutoff are collected together and then two loops sum force and energy contributions for each pair of interacting groups. The structure of these routines are the same so we only show pseudo-code for one of them in figure 2.

Once the energies and full force vectors have been assembled, the system is advanced to the next time step by integrating the equations of motion. This uses a leap frog integrator [13]. Optionally, the program enforces a set of input constraints with the shake algorithm [13] and if necessary scales the velocities to keep the system temperature within the preset simulation value.

The parallel WESDYN program

We called the parallel version of WESDYN, pWESDYN. Out first decision was which of the many parallel MD algorithms to use. It is well known that for the best performance on massively parallel machines, a method based on a spatial domain decomposition [5] is required. These algorithms, however, are difficult to program and requires extensive modification to the sequential version of the program. This violates our goals to maintain a single version of the program's source code as well as the need for simple parallelization.

Parallel MD algorithms based on atom (or atom group) decompositions, however, are simple to program requiring few modifications to the sequential code. These methods do not scale well for large numbers of nodes, but with our principle target being workstation clusters, this was not a problem.

When parallelizing programs, it is important to focus on the computational bottlenecks. In molecular dynamics, the computational bottlenecks are the non-bonded energy computations which usually consume more than 90% of the total elapsed compute time. Generation of the non-bonded interaction lists is also compute intensive (complexity $O(N^2)$), but since this is only carried out occasionally we did not parallelize this operation.

To parallelize the non-bonded computation, we used a technique known as loop splitting [9]. In a loop splitting algorithm, the program is organized as an SPMD code with the principle data structures (forces and coordinates) replicated on each node. This lets one express the parallelism by assigning loop-iterations to different nodes of the computer. Rather than fixing a particular partitioning into the code, we used an *owner-computes filter*. In this method, an `if-statement` at the top of the loop tests to see if the node *owns* that particular iteration of the loop. This if-statement filters the loop iterations so different nodes compute different iterations. The advantage of this method is the simplicity with which it can be added to code. It also provides a great deal of flexibility to the programmer exploring different load balancing algorithms.

The loop splitting method can be applied to each time-consuming part of the program. We choose, however, to only parallelize the non-bonded energy terms and then redundantly update the sequential portions of the code. This seems wasteful, but given the slow performance of local area networks, it was deemed appropriate since it can be faster to compute these terms than to communicate them. For larger systems (50,000 atoms and larger) and larger numbers of nodes, it is important to parallelize the routines that generate the non-bonded lists. We have only recently done this and will address this in a later paper.

Once the basic algorithm was selected, we had to select a programming environment that would support parallel computing on a network of workstations as well as more traditional parallel computers. We used two programming environments in this project: Fortran-Linda [11] and TCGMSG [6].

While a full description of Linda is well beyond the scope of this paper, we need to understand just enough of Linda to follow the pseudo code for pWES-DYN. Linda [7] is based on a virtual shared memory through which all interaction between processes is managed. These memory operations are added to a sequential language to create a hybrid parallel programming language. For example, in pWESDYN, the combination of Fortran and Linda was used [11].

Linda consists of four basic operations. The first operation is called eval(). This operation spawns a process to compute the value returned by a user provided function. When the function is complete, the value is placed into Linda's shared memory. The second operation, called out(), does the same thing as eval() except out() doesn't spawn a new process to compute the values to deposit into Linda's shared memory. In other words, eval() is parallel while out() is sequential.

The other two Linda operations are used to extract information from the shared memory. If some process wishes to fetch some data and remove it so no other process can grab the item, the in() instruction is used. Finally, the rd() operation grabs the data but leaves a copy behind for other processes.

The last concept to cover is how the programmer specifies which items to access in Linda's shared memory. Items in Linda's shared memory are accessed by association – not by address. In other words, the Linda programmer describes the data they are looking for and the system returns the first item in shared memory that matches that description. If nothing matches the description, the Linda operation blocks (i.e. waits) until such an item exists. This description, called a *template*, plays a key role in Linda.

Templates are defined in terms of the number, types and values of the arguments to in() or rd(). In addition, it is frequently necessary to indicate a location in the program's memory (not Linda's memory) to hold items pulled out of Linda's shared memory. This is done by specifying *placeholders* in the template definition. The symbol used for the placeholder is a "?" preceding the variable that will hold the item in question.

A few simple examples will clarify this discussion. To create items in Linda's memory, we would use the commands:

```
out  ('I am a Linda memory item', 10)
eval (' this is the output of function g', g() )
out  (10, 2.345)
```

To grab the last item in this set, one would use the Linda operation:

```
in (10, ?x)
```

which indicates that the in() operation wants to access a two field item from Linda's shared memory with the first field containing an integer of value 10 and the second filed containing an item with the same type as x. At the conclusion of this operation, the variable x would contain the value from the second field – in this case 2.345. If the object being fetched from Linda's shared memory is an array, the in() or rd() statement uses a colon and a variable to hold the length of the array that was fetched. For example, the statement:

```
in("an array", 5, ?y:n)
```

will fetch an item from Linda's shared memory with the first field containing the string "an array", the second field an integer of value 5, and an array with the same type as y in the last field. At the conclusion of this operation, the array y will hold a number of elements given by the value of the integer n.

At this point, we have covered enough Linda to read the pseudo code used in this chapter. For a more detailed introduction to Linda, see the book by Carriero and Gelernter [7].

The Linda version of pWESDYN was structured around two processes: a *master* and a *worker*. Pseudo code for the master is given in figure 3. The master sets up the calculation by spawning a process on the other nodes where the process will execute the worker-function. This function takes two arguments to provide a worker ID and a value for the total number of workers.

The master then waits for the workers to finish. This is an example of the use of Linda's memory to provide synchronization. The master process blocks until an item in Linda's memory exists that contains two fields: the string 'worker process' and the return value from the function worker(). The master will execute one in() for each eval()'ed process thereby assuring that it waits until everyone is done.

In this version of the code, the master just starts up the job, and waits for indication of completion. There is no reason the master couldn't transform into a

```
program pWESDYN
    number_of_workers = READ_NUMBER_OF_WORKERS
    do i=1,number_of_workers      !create workers
       eval('worker process', worker(i,number_of_workers))
    end do
    do i=1,number_of_workers      !Confirm the completion
       in('worker process', ?k)
    end do
end pWESDYN
```

Figure 3: Pseudo-code for pWESDYN master code

```
integer function worker(my_ID,number_of_workers)
    CALL INITIALIZE_OWNER_COMPUTES_FILTER ()
    CALL WESDYN()
    return
end worker
```

Figure 4: Pseudo-code for pWESDYN worker code

worker after the eval-loop. We choose not to do this in order to make performance evaluation easier.

The worker code is presented in Figure 4. This is also a very simple routine. The worker sets up the owner computes filter by calling OWNER_COMPUTES_SETUP() (which will be explained later), and then calls the WESDYN subroutine.

At this point, the two routines have done nothing but setup the parallel program. The work carried out by the worker is within the routine, WESDYN(). We present pseudo-code for the parallel WESDYN() routine in Figure 5. Notice that all the parallelism is bracketed by #ifdef-statements which lets us maintain the parallel code within the same source code as the sequential program. Also notice that the difference between the sequential and parallel versions of WESDYN() are minimal.

There is no substantial difference between the sequential and parallel versions of WESDYN(). The only exception is within WESDYN() all output is bracketed by #ifdef PARALLEL statements to select code that directs only one node to create and write the output files.

The parallelism is buried within the non-bonded routines. Hence, the only other changes required for the parallel version of the program are in each of the three non-bonded energy routines. In this paper, we will only show the changes within NBSTST() since the other two are of nearly identical structure.

In Figure 6 we show pseudo-code for NBSTST(). Note that NBSTST() was changed in only three locations – otherwise the code is identical to the sequential version. Basically, the outermost loop over charge groups has been amended at the top of the loop with an owner-computers filter. This mechanism is simple to implement and lets us conduct future research on load balancing algorithms.

Since different loops are executed by different nodes, the forces are fragmented across the workers. The global sum after the loops, reconstitutes the fragmented force vector into a single vector and assures that a copy of this summed vector resides on each node. In the call to the global sum:

```
CALL GDSUM(F,3*N+6,WORK)
```

```
#ifdef PARALLEL
subroutine WESDYN()
#else
program WESDYN
#endif
      PROCESS_USER_COMMANDS
      READ_MOLECULAR_TOPOLOGY_FILE
      READ_CARTESIAN_COORDINATES
      READ_INITIAL_VELOCITIES
      MINIMIZE_ENERGY_OF_INITIAL_STRUCTURE
      do i = 1,number_of_steps
         if (TIME_TO_UPDATE_NON_BONDED_NEIGHBOR_LIST) then
                 EVALUATE_NON_BONDED_NEIGHBOR_LIST
         endif
         CALCULATE_INTERNAL_COORDINATE_ENERGIES_AND_FORCES
         CALCULATE_GEOMETRIC_CENTERS_OF_CHARGE_GROUPS
         CALL_NBSTST  ! evaluate solute-solute energies and forces
         CALL_NBSTSV  ! evaluate solute-solvent energies and forces
         CALL_NBSVSV  ! evaluate solvent-solvent energies and forces

         INTEGRATE_EQUATIONS_OF_MOTION
         if (SHAKE_REQUESTED) then
            APPLY_SHAKE
         endif
         if (TEMPERATURE_OFF_LIMITS) then
            RESCALE_VELOCITIES
         endif
      end do
#ifdef PARALLEL
      return
#endif
end WESDYN
```

Figure 5: Pseudo-code for WESDYN() subroutine. Except for trivial changes, this is identical to the pseudo-code for the sequential WESDYN program

```
subroutine NBSTST
      do i = 1,number_of_charge_groups
#ifdef PARALLEL
        if (mine(i)) then
#endif
          number_of_interacting_groups =
          COLLECT_ALL_GROUPS_WITHIN_CUTOFF
          do j = 1,number_of_interacting_groups
            SWITCH =
              CALCULATE_SWITCHING_FUNCTION(group(i),group(j))
            do k = 1,number_of_atoms(group(i))
              do l = 1,number_of_atoms(group(j))
                E_SOLUTE_SOLUTE = E_SOLUTE_SOLUTE + SWITCH *
                        ENERGY(atom(k,group(i)),atom(l,group(j)))
                F_PAIR = FORCE(atom(k,group(i)),atom(l,group(j)))
                F(atom(k,group(i))) = F(atom(k,group(i)))
                        + SWITCH * F_PAIR
                F(atom(l,group(j))) = F(atom(l,group(j)))
                        - SWITCH * F_PAIR
              end do
            end do
          end do
#ifdef PARALLEL
        endif
#endif
      end do
#ifdef PARALLEL
      CALL GDSUM(F,3*N+6,WORK)     ! combine results
#endif
      end NBSTST
```

Figure 6: Pseudo-code for the parallel NBSTST subroutine.

```
subroutine GDSUM(F,N,WORK)
    my_ID = mynode()
    if (my_ID .eq. GDSUM_NODE) then    !The node designated to
        do i=1,number_of_workers-1     !perform global sum collects
            in('gdsum data',?work:len3) !energy-force array from all
            do j=1,N                   !other nodes and accumulates
                F(j) = F(j) + work(j)  !them.
            end do
        end do
        do i=1,number_of_workers-1     !Send out copies of summed up
            out('gdsum data_answers',X:N) !energy-force array to all
        end do                         !other workers.
    else
        out('gdsum data',X:N)          !Other workers initially send
                                       !out their energy-force arrays
        in('gdsum data_answers',?X:len3)!wait for the sum to come
                                       !back
    endif
```

Figure 7: Pseudo-code for a global sum.

F is an array containing the six energy components plus the 3*N force vector and
WORK is a temporary work array. The operation of a global sum is to carry out an
element-wise sum across the elements of a vector and to place an identical copy
of the result on each node of the parallel computer. In Figure 7 is a routine to
carry out the global sum operation. The algorithm used here is primitive and
far better methods are known [8]. We tested this method against more optimal
methods that use traversal of a balanced binary tree to guide the global sum. We
found that on small workstation clusters, our primitive method was competitive
(and in some cases even faster). This of course would not be the case for parallel
computers with large numbers of nodes.

NBSVSV() was changed along the same lines as NBSTST(). NBSVST() was modified along slightly different lines. In this case, it did not include the call to the
global sum and the fragmented force vector was passed onto NBSVSV(). This saved
one global sum operation.

The last component of the program we need to explain is the owner computes filter in figure 8. An array (called loop_ownership is initialized to zero
and stored as a global variable (i.e. stored within a common block). This is
a two state filter with zero representing the case where some other node owns
the iteration and one represents ownership of the iteration. In the subroutine
INITIALIZE_OWNER_COMPUTES_FILTER(), we show the cyclic decomposition used
in many MD codes. Finally, the logical function MINE() accesses the global array

```
subroutine INITIALIZE_OWNER_COMPUTES_FILTER()
     common/filter/loop_ownership (MAXATOMS)
     do i=1, MAXATOMS
        loop_ownership(i) = 0
     end do
c
     do i=my_ID+1,MAXATOMS,number_of_workers
        loop_ownership(i) = 1
     end do
end INITIALIZE_OWNER_COMPUTES_FILTER

logical function MINE (i)
     common/filter/ loop_ownership (MAXATOMS)
     if(loop_ownership(i) .eq. 1) then
        return .true.
     else
        return .false.
     endif
end MINE
```

Figure 8: Pseudo-code for the code to setup the owner compute filter.

to return the state for the indicated iteration.

This approach to distributing loop iterations may at first seem unnecessarily complex. However, just by changing one of two simple routines, we can experiment with different load balancing strategies. This will be the focus of the next phase of our research.

pWESDYN program: the TCGMSG version

Linda was the first programming environment we used to code pWESDYN. We wanted to use an additional programming environment for two reasons. First, Linda is a commercial product and is not available on every system we wished to utilize. By creating a version of pWESDYN based on a public domain programming environment, we could move the programming environment anywhere we wished. Second, we wanted to verify that the observed performance was due to the algorithm and not the Linda programming environment.

The additional programming environment we selected was TCGMSG [6]. TCGMSG is a coordination library consisting of message passing constructs and global communication operations. The global operations proved particularly convenient and

```
program pWESDYN
      call PBEGINF        ! initialize TCGMSG
      call INITIALIZE_OWNER_COMPUTES_FILTER()
      call WESDYN
      call PEND            ! shut down TCGMSG
end pWESDYN
```

Figure 9: Pseudo-code for TCGMSG version of pWESDYN.

saved us having to develop these routines ourselves.

Generation of the TCGMSG version of the program was almost trivial once the Linda version existed. This simple program is shown in Figure 9. The TCGMSG runtime environment handles process startup. Therefore, the program is rigorously an SPMD (Single Program Multiple Data) program in which each node runs the identical program. The only change we had to make was to provide an interface between our definition of the global sum and the one that comes with TCGMSG.

Results

We have studied the performance of pWESDYN on a wide range of MIMD systems. The code was portable from workstation clusters, to shared memory multiprocessors, to distributed memory MIMD supercomputers. Of these systems, however, we have only fully analyzed the results from workstation clusters. Hence, in this paper we will only discuss the workstation cluster results and save the other systems for a future paper.

The cluster in this study was a network of RS/6000 560 workstations with 128 Mb of random access memory on each workstation. These were connected by an ethernet Local Area Network. This cluster was a shared resource, but it was managed so dedicated access could be reserved for benchmarking.

To benchmark pWESDYN, we worked with a dodecamer sequence of DNA known as the B80 canonical structure of the Drew, Dickerson sequence [1], d(CGCGAATTCGCG). We did not use counterions to balance the backbone charges of the DNA. Rather, in accordance with Manning's counterion condensation theory [10], the phosphate ions were reduced to −0.25.

The DNA dodecamer contained 542 atoms. To provide a 12 Angstrum solvent shell around the DNA molecule, we added 3580 SPC waters in a hexagonal box. The system was equilibrated prior to MD by long Monte Carlo simulation on just the water molecules. Finally, the non-bonded interactions were slowly switched

Table 1: Wall clock times, Total CPU times, and CPU times within non-bond routines for pWESDYN on the RS/6000 cluster. This particular calculation was for 50 energy minimization steps.

numb. of	TCGMSG			Linda		
nodes	Wall	Cpu(Tot)	Cpu(NB)	Wall	CPU(Tot)	CPU(NB)
1	1349.3	1341.68	1281.28	1348.2	1340.0	1280.0
2	777.0	740.3	679.3	799.7	736.5	671.7
3	690.0	543.2	482.0	624.7	532.0	466.4
4	637.5	446.5	385.4	588.9	444.2	375.2
5	665.3	380.9	320.1			
6	652.1	339.5	278.5			

All times are in seconds.

from 7.5 to 11.5 angstroms.

In Table 1 we report on the times for computing 50 energy minimization steps for the benchmark system. The energy minimization routines call the same parallelized non-bonded energy routines, so this benchmark problem provides a usefully sized problem for evaluating pWESDYN. This data shows that the TCGMSG program runs at the same speed as the Fortran–Linda program for small numbers of nodes but by three nodes, it is on the order of 10% slower. We believe this is due to the the TCGMSG global communication routines. The TCGMSG routines dynamically allocate memory as needed while the Linda routines reduced runtime memory overhead by using a static work array.

The maximum speedup in Table 1 is 2.3 which occurs for Linda at 4 nodes. The speedup in terms of the CPU time, however, is 3.0 at 4 nodes and (in the case of TCGMSG) continues to improve with more nodes. This discrepancy between CPU and Wall times is caused by time spent with additional communication as well as managing additional ethernet collisions as more nodes are added. This impacts the elapsed wall-clock time but is not accounted to the process CPU time.

At first, a maximum speedup of 2.3 seems quite disappointing. This is on par, however, with other molecular dynamics programs running on ethernet clusters. For example, in [12] a maximum speedup of 2.5 was reported for their ethernet cluster of RS/6000 560 workstations. This is particularly noteworthy in light of two facts. First, every facet of the program used in [12] was parallelized – not just the non-bonded routines. Second, the system studied in [12] was much larger (greater than 14,000 atoms) than our system (3,922 atoms). Given that both communication and computation scale as O(N) in a cutoff based MD computation, one might predict that system size does not impact efficiency. In the complexity analysis, however, the term multipling N is much larger for computation than for communication so there is far more computation to parallelize for larger systems.

Table 2: Wall clock times, Total CPU times, and CPU times within non-bond routines for pWESDYN on the RS/6000 cluster. The calculation was for .35ps of dynamics, 2 step minimization, and with I/O every .05 ps.

numb. of	Linda		
nodes	Wall	Cpu(Tot)	Cpu(NB)
1	4853	4749	4514
2	2879	2602	2332
3	2262	1848	1623
4	2073	1476	1250

All times are in seconds.

(For a recent example of this effect, see [14]). Given these two effects that favor the work in [12], our speedup of 2.3 is actually quite respectable.

In Table 2 we show the results for a .35 picosecond (350 steps) MD simulation with 2 minimization steps prior to the simulation. This test stresses the program in a full production mode with I/O occurring every .05 picoseconds (50 steps). Notice that the qualitative trends seen in the minimization benchmark are reproduced for this MD benchmark. This is the expected result given the portions of the program that was parallelized. We use this result to justify basing our comparisons on the simpler, energy minimization calculations.

Conclusion

In this project, we investigated some of the software engineering issues behind parallel molecular dynamics programs. To this end, we parallelized a major molecular dynamics program called WESDYN [3]. For the parallelization, we set the following guidelines:

- Design the parallel algorithm so any changes to the sequential code will be simple and well isolated.

- Implement the changes so the parallel and sequential programs can be maintained within within the same source code.

- Support portable execution of the parallel code among different MIMD parallel systems.

Performance of the parallel program was important, but of a secondary concern relative to the issues of quality software engineering. In addition, the primary target was modest numbers of clustered workstations, though we wanted to write portable code that could easily move to moderate sized MIMD parallel computers as well.

On the first two points, we clearly succeeded. The parallel algorithm described in this paper was simple to implement and required so few changes that a single source with `ifdefs` was able to contain both the sequential and parallel programs.

The third point – portability to other MIMD computers – will be the focus of the next phase of this work. While the results are too preliminary to analyze in any detail, we can briefly relay some preliminary results. The code was portable in that correct answers were produced on a range of MIMD systems. However, by having each node separately open and read data files, the code did not perform well on distributed memory systems. To deal with this effect, we need to add parallel I/O to our methodology. The common approach is to have an `#ifdef` in the code so for parallel systems, only one node accesses the file and then broadcasts the results to the other nodes. The problem is that within fortran, there is no easy way to do this without carefully breaking down the messages based on the size and type of the data. This requires an intimate level of knowledge with the code and is therefore undesirable with dusty deck codes. We hope to find a way to semi-automatically apply this transformation.

References

[1] S. Arnott, R. Chandrasekaran, D.L. Birdsall, A.G.W. Leslie, and R.L. Ratliffe, *Nature, Vol 283*, p. 743, 1980.

[2] W.F. van Gunsteren and H.J.C. Berendsen, GROMOS88: Groningen Molecular Simulation System, University of Groningen, The Netherlands, 1988.

[3] G. Ravishanker, WESDYN 2.0, Wesleyan University, 1993.

[4] S. Swaminathan and D.L. Beveridge, WESDYN 1.0, Wesleyan University 1988.

[5] S. Plimpton, "Fast Parallel Algorithms for Short-Range Molecular Dynamics," Sandia Technical Report, SAND91-1144, 1993.

[6] R. J. Harrison, "Portable Tools and Applications for Parallel Computers," *International Journal of Quantum Chemistry, Vol 40*, 847-863, 1991.

[7] N. Carriero and D. Gelernter, *How to Write Parallel Programs: A First Course.* Cambridge: MIT Press, 1990.

[8] R.A. van de Geijn, "Efficient global combine operations." *Proceedings Sixth Distributed Memory Computing Conference*, p. 291, IEEE Computer Society Press, 1991.

[9] T. G. Mattson, "Scientific Computation," *Handbook of Parallel and Distributed Computing*, ed. A. Y. Zomaya, McGraw Hill, 1995.

[10] G.S. Manning, *Quart. Rev. Biophys. Vol 11*, 179, 1978.

[11] Scientific Computing Associates, Inc., " Fortran-Linda User Guide and Reference Manual," New Haven, CT, 1993.

[12] B.R. Brooks and M. Hodoscek, "Parallelization of CHARMM for MIMD machines," *Chemical Design Automation News*, vol 7, p. 16, 1992.

[13] For a general description of MD methodology, including SHAKE, see van Gunsteren, W.F., and Berendsen, H.J.C., *Angew.Chem.Int.Ed.Engl.*, *29*, p. 992-1023, 1990.

[14] J.A. Lupo, "Benchmarking UHGROMOS," *Proceedings of the 28 Hawaii International Conference on System Sciences*, IEEE Computer Society Press, 1995.

RECEIVED November 15, 1994

Chapter 11

Advanced Algorithms for Molecular Dynamics Simulation

The Program PMD

Andreas Windemuth

Department of Biochemistry and Molecular Biophysics
and Center for Biomolecular Simulation, Columbia University,
630 West 168th Street, New York, NY 10032

A collection of algorithms is presented to allow for the efficient computation of the dynamics of large systems of macromolecules and solvent. Application of the Fast Multipole Algorithm coupled with the Distance Class Algorithm, a multiple timestep method, permits the evaluation of unlimited long-range interaction at a cost lower than that of conventional cutoff calculations. A new method for the calculation of analytical surface areas and derivatives, the Circle Intersection Method (CIM), is also described. The CIM is at least 2–3 times faster than existing exact analytic methods. All methods described in this paper are designed to be scalably parallel, meaning that resource requirements grow at most linearly with the size of the system and are inversely proportional to the number of processing nodes for sufficiently large systems. The experimental program PMD, designed to implement these methods, is described and plans for its future development with emphasis on advanced solvent modeling is outlined. PMD is made freely available at this point to provide a flexible testing ground for advanced algorithms to all interested researchers.

Molecular dynamics simulation has become an important tool in computational chemistry, particularly for the modelling of biological macromolecules (1–4). The method requires the evaluation of forces acting on each atom in the simulated system, which often contains a large number of solvent molecules as well as complex heterogeneous macromolecules such as proteins. The rapid increase in computational capacities has made it possible in recent years to perform simulations of large solvated systems with tens of thousands of atoms, and even much bigger systems are becoming feasible.

0097–6156/95/0592–0151$12.00/0
© 1995 American Chemical Society

A sizable choice of programs are available to do molecular dynamics simulations, most of which reach back more than a decade and were originally designed to accomodate a few hundred or a thousand atoms (5–7). The traditional way to deal with the long range Coulomb interactions is to neglect them beyond a certain distance, thereby causing significant errors (8). The programs were optimized to run fast on the vector supercomputers that were state of the art at the time, and it is not clear how well they can be adapted to the upcoming generation of parallel computers, most of which are based on superscalar microprocessors connected by a message passing communications network.

The present paper will attempt to address these problems by presenting a collection of advanced algorithms embodied in the intrinsically parallel and distributed program PMD. The goal of PMD is to provide infinitely scalable parallel molecular dynamics computation, in the sense that any size of simulation can be performed in a given time if enough processing nodes are available. The basic requirement for infinite scaling is that computation time, memory usage and communication bandwidth all increase at most linearly with the number of atoms and decrease inversely proportional to the number of available processing nodes. This requires distributed storage of atom parameters and coordinates, and the decomposition of the system must be according to spatial regions, in order to eliminate non-local communication.

Another design principle of PMD is that long range Coulomb forces are not to be neglected. This is achieved by the adoption of the parallel fast multipole algorithm (PFMA) (9), which performs an arbitrarily accurate calculation of the full Coulomb interactions in a time of order $O(N)$, with N being the number of atoms in the simulated system. To make the computation time for the full interactions comparable to those of cut-off calculations, the Distance Class Algorithm, a simplified version of the Generalized Verlet algorithm (10), is provided in PMD. This method, similar to an earlier method by Teleman and Jönsson (11), separates the slowly changing long range Coulomb interactions from the short range interactions and keeps their forces constant, to be recalculated at regular intervals during the simulation.

With regards to future development, PMD is intended to be the test bed for other advanced algorithms, particularly implicit solvent models and continuum elctrostatics methods, as well as accelerated and directed simulation techniques to be used in the study of ligand binding and protein folding. Many implicit solvent models, including the continuum electrostatics methods, require a definition of the solvent accessible or molecular surface. Moreover, if these models are to be used with dynamics simulation or minimization, the derivatives of the surface area with respect to the atom coordinates have to be available. To provide the basis for future solvent modelling, PMD currently incorporates a novel, very efficient algorithm, the Circle Intersection Method (CIM), to analytically calculate accessible surface areas of macromolecules and their derivatives.

Data Distribution

The question of how to distribute the atom coordinates and force field parameters across processing nodes is crucial for the design of a parallel algorithm. The simplest way to parallelize MD simulations is to distribute all parameters and coordinates to all processing nodes, calculate a different subset of the force field on each node, and globally combine and add the resulting partial forces. The advantage of this full distribution method is that it requires a minimum amount of changes in existing code, and load balancing is relatively easy to implement. Also, no special considerations have to be given to non-local interactions, such as the long range Coulomb interactions. The scaling of the method is $O(N/P)$ for time, $O(N)$ for storage per node and $O(N)$ for communication bandwidth per node. The full distribution method is useful for small and medium sized systems, and on machines with a small number of nodes, each containing sufficient memory to hold the full system. A fast communication network is necessary, due to the need for global communication. Obviously this method will quickly reach its limits when large systems and massively parallel machines are involved.

A better way of data distribution is the force decomposition method (12, 13). The atoms are divided into $\sqrt{P}$ classes, and each node stores the coordinates and parameters of the atoms in two such classes, pairing different classes on each node. This method is more scalable than the full distribution method, but it is also more difficult to implement. The scaling is $O(N/P)$ for time, $O(N/P)$ for storage per node and $O(N/\sqrt{P})$ for communication bandwidth per node. Thus, communication is not scalable, albeit much more so than with the full distribution method.

In order to obtain a fully scalable algorithm, a spatial decomposition method has to be used, which takes advantage of the spatial locality of the short range interactions. The simulation volume is divided into P separate regions, and each processing node is assigned all atoms in one of these regions. Communication of coordinates and forces takes place only between nodes whose regions are within the maximum interaction distance of each other. The scaling of this method is $O(N/P)$ for time, $O(N/P)$ for storage per node, and $O((N/P)^{\frac{2}{3}})$ for communication bandwidth per node. Special consideration has to be given to the long range interactions in the spatial decomposition method, since only interactions within a certain distance are covered by the communication scheme. By adding an additional communication step to the PFMA, in which the values of multipoles are exchanged, the communication bandwith per node of the long range interactions can be made nearly scalable, meaning that its complexity will be $O((N \ln N)/P)$.

In PMD, a flexible spatial decomposition method, called *Voronoi Decomposition*(VD), is used to provide load balancing and allow for arbitrarily shaped molecules. Each processing node is assigned a position in the simulation region, and atoms are assigned to whichever node they are closest to. This leads to a decomposition of space into Voronoi polyhedra whose boundaries are defined by the othogonal midplanes along the distance between each pair of nodes, as illustrated in figure 1.

There are several advantages to using Voronoi decomposition as opposed to a more conventional cubic grid decomposition. The most important advantage is that by shifting the positions assigned to the nodes, the size of the regions can be varied to provide both static and dynamic load balancing. Load balancing is needed when the atom density is non-uniform throughout the simulation region, as well as when the processing nodes run at different speeds or are differently loaded, a common situation in workstation clusters.

To determine the directions in which the node positions need to be shifted,

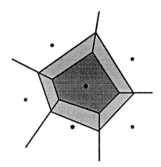

Figure 1: Spatial decomposition using Voronoi polyedra. The dark gray region contains the atoms assigned to a particular node, the light gray region contains interacting atoms for which coordinates and forces have to be communicated to and from neighboring nodes. Only two dimensions are shown, actual decomposition is in three dimensions.

each node distributes the actual time t_i spent on its part of the potential evaluation to all neighbors. Each node then determines the average t_{avg} of the load among itself and its neighbors. The node positions p_i are then periodically updated according to

$$p_i' = p_i + \omega \sum_{j \in N_i} \left(\frac{t_j - t_i}{2t_{avg}} + \frac{1}{|p_j - p_i|^6} \right) (p_j - p_i), \tag{1}$$

where the sum is performed over all interacting neighbors of node i. The first term in the sum pulls overloaded nodes towards other nodes, thereby reducing the volume of their Voronoi polyhedron. The second term keeps node centers at a distance from each other to avoid singularities. The particular form of these terms is highly empirical and has not been fully explored. The factor ω is adjusted to provide optimum convergence of the procedure.

For the calculation of all short range interactions, each node stores the coordinates and parameters of the atoms within its domain, called local atoms, as well as copies of coordinates and parameters from atoms in neighboring domains, called ghost atoms. Only ghost atoms that are within the maximum interaction distance d_c from the boundary of the Voronoi polyhedron are actually stored, which, in the limit of large domains, makes the cross-boundary communication a surface effect and leads to the $O((N/P)^{\frac{2}{3}})$ scaling of communications.

In order to avoid complicating the development of force fields and simulation methods, the actual form of decomposition is hidden behind a Data Distribution Interface (DDI), keeping the overhead associated with developing parallel code to a minimum and permitting transparent change of the underlying decomposition algorithm without affecting the Algorithms built on it. For example, dynamic load balancing is completely transparent and can be achieved simply by calling a balancing routine at appropriate times during the simulation.

Molecular Structure and Potential Function

PMD was built to implement the widely used CHARMM force field, the form of which will be summarized in this section. Molecular Dynamics simulation is based on the numeric integration of Newton's equations of motion:

$$m_i \ddot{\mathbf{x}}_i = \nabla_i U(\mathbf{x}_1, \dots, \mathbf{x}_N). \tag{2}$$

The vectors $\mathbf{x}_i$ denote the positions of all atoms, with i ranging between 1 and the total number of atoms N. The m_i are the atomic masses and $U(\mathbf{x}_1, \dots, \mathbf{x}_N)$ is the full potential function representing all interactions between the atoms. The potential is composed of bond and nonbond interactions:

$$U = U_r + U_\theta + U_\phi + U_\omega + U_{\text{vdW}} + U_{\text{Coulomb}}. \tag{3}$$

The bond interactions U_r, U_θ, U_ϕ, and U_ω represent the forces between atoms generated by their chemical bonding, and are approximated by the following expressions:

$$U_r = \sum_{n=1}^{N_r} k_{r,n} (r_{i_n j_n} - r_{0,n})^2 \tag{4}$$

$$U_\theta = \sum_{n=1}^{N_\theta} k_{\theta,n} (\theta_{i_n j_n k_n} - \theta_{0,n})^2 \tag{5}$$

$$U_\phi = \sum_{n=1}^{N_\phi} k_{\phi,n} [1 - \cos n_n (\phi_{i_n j_n k_n l_n} - \phi_{0,n})] \tag{6}$$

$$U_\omega = \sum_{n=1}^{N_\omega} k_{\omega,n} (\omega_{i_n j_n k_n l_n} - \omega_{0,n})^2 \tag{7}$$

U_r is composed of an harmonic stretching force between two atoms i_n and j_n for each one of N_r bonds. The quantities $k_{r,n}$ and $r_{0,n}$ are parameters to the force field and depend on the properties of the atoms i_n and j_n as expressed by their atom types. The other terms are similarly defined to describe harmonic bending of the angle between two atoms, a periodic potential for torsions involving four atoms and an harmonic improper torsion potential to provide planarity constraints in ring structures as well as tetrahedral centers. The potential depends on the atom coordinates through the bond lengths $r_{ij} = |\mathbf{x}_i - \mathbf{x}_j|$, the bond angles θ_{ijk}, and the proper and improper torsion angles ϕ_{ijkl} and ω_{ijkl} (6).

At the center of the force field parametrization are the atom types. Atoms are assigned one of several types according to their elemental nature, hybridization state and other relevant chemical properties. Bonds, angles, dihedrals and improper torsions are then assigned to pairs, triples and quadruples of atoms according to the chemical structure of the molecule. The parameters $(k_{r,n}, r_{0,n}, k_{\theta,n}, \theta_{0,n}, k_{\phi,n}, n_n, \phi_{0,n}, k_{\omega,n}, \omega_{0,n})$ are tabulated for each possible combination of atom types. They are usually obtained from fitting to experimental results or from quantum chemical calculations.

In order to simplify the assignment of atom types and bond topology, the assignment is done for small groups of atoms called *residues*, which can be anything from a water molecule to a chromophore. The most common residues are the amino acids that proteins are composed of. Larger structures such as proteins are assembled from the residues by means of *patches*, i.e. rules for connecting together residues. Patches specify the modifications and additions that have to be applied to the participating residues in order to describe the mechanics of the combined system correctly. One example for a patch in proteins is one describing peptide bonds, another is one for disulfide bonds.

For proteins, all that is needed to assign the bond energy parameters is the sequence of residues and a specification which cysteine residues should be involved in a disulfide bond. PMD then automatically applies the appropriate peptide and disulfide patches to generate the complete molecular structure. Solvent molecules can be specified as residues with a replication number specifying the number of solvent molecules that are present. CHARMM format residue topology files and parameter files are read by PMD to define the residues and assign the parameter values.

The remaining terms in equation 3 describe the nonbond potential. They are composed of van der Waals interactions described by the Lennard-Jones potential

$$U_{\text{vdW}} = \sum_{i,j} \left(\frac{A_{ij}}{r_{ij}^{12}} - \frac{B_{ij}}{r_{ij}^6} \right), \tag{8}$$

and Coulomb interactions

$$U_{\text{Coulomb}} = \sum_{i,j} \frac{q_i q_j}{\epsilon r_{ij}}, \tag{9}$$

and depend on the atom coordinates solely through the interatomic distances $r_{ij} = |\mathbf{x}_i - \mathbf{x}_j|$. The parameters A_{ij} and B_{ij} are tabulated for all combinations of atom types. They are usually derived from per-atom parameters ϵ_i and σ_i in the following way:

$$A_{ij} = 4 \left(\frac{\sigma_i + \sigma_j}{2} \right)^{12} \sqrt{\epsilon_i \epsilon_j}, \qquad B_{ij} = 4 \left(\frac{\sigma_i + \sigma_j}{2} \right)^6 \sqrt{\epsilon_i \epsilon_j}, \tag{10}$$

but there may be exceptions where A_{ij} and B_{ij} may be specifically assigned for certain pairs of atom types.

The partial charges q_i are specified independently of atom type in the residue topology file and are chosen to approximate the actual charge density of the residues as derived from quantum chemical calculations. The dielectric constant is normally $\epsilon = 1$, but may be set to something larger to account for electronic polarization effects.

The bond potential does not require much time to calculate, since the number of bonds, angles and dihedrals is on the order of the number of atoms. The nonbond potential, however, is defined as a sum over all pairs of atoms, i.e. a sum with $N(N-1)/2$ terms. For large systems, full calculation of this sum is prohibitive. The Lennard-Jones potential (equation 8) used to describe the van der Waals interactions is short range, i.e. its strength diminishes quickly outside a certain range. Thus, it can be calculated in good approximation as a sum over all pairs within a maximum distance d_c, the *cutoff* distance.

The Coulomb potential (equation 9), however, is long range, i.e. it does not decrease with distance enough to make up for the increasing number of atoms at that distance. It has traditionally also been calculated using a cutoff, but this has been known to be a bad approximation for macromolecular systems (8). There are often extended charge imbalances in proteins such as helix dipoles, charged side chains, and dipolar membranes, the electrostatic properties of which may contribute considerably to the structure and function of the system. It is precisely to address this problem of the long range nature of the Coulomb interactions, that the Fast Multipole Algorithm is used in PMD, which will be the focus of the following section.

The Fast Multipole Algorithm

In order to efficiently compute the full Coulombic interactions in large molecular systems, the Fast Multipole Algorithm as proposed by Greengard and Rokhlin (14, 15) and later implemented by Leathrum and Board (9, 16) is used in PMD. The algorithm allows the calculation of the full Coulomb interaction to a given numerical precision in a time proportional to the number of atoms. It thus satisfies the requirement of scalability. The implementation by Leathrum and Board, called the Parallel Fast Multiple Algorithm(PFMA) is very efficient, due mostly to the precomputation of coefficients and the use of spherical harmonics, and it was designed to run on parallel machines, making it very suitable for use in PMD.

A similar algorithm, the Cell Multipole Method, was developed independently by Ding and Goddard (17). Speedups similar to the FMA have been observed with the CMM, and tested for systems with up to a million atoms. The CMM is based on the physically more intuitive cartesian representation of multipoles, while the FMA is based on the mathematically more appropriate spherical harmonics functions. Since cartesian multipoles of higher order are more cumbersome to implement than the corresponding spherical harmonics, the accuracy of the CMM as reported by Goddard is restricted to the octupole level, corresponding to $p = 3$ in the FMA terminology below, while the accuracy of the FMA is limited only by memory and CPU time requirements. No parallel implementation of the CMM has been reported.

Some novel improvements were made to the PFMA code upon integration into PMD, including a pruning of chargeless volumes to avoid unnecessary computation and storage, and a task distribution based scheme for load balancing.

The Fast Multipole Algorithm is based on the expansion of the Coulomb potential of a bounded charge distribution in multipoles

$$\Phi(\mathbf{x}) = 4\pi \sum_{l,m} \frac{M_{lm} Y_{lm}(\theta, \phi)}{(2l + 1) r^{l+1}}, \tag{11}$$

where (r, θ, ϕ) are the spherical coordinates associated with the cartesian coordinate vector $\mathbf{x}$, and $Y_{lm}(\theta, \phi)$ are the complex valued spherical harmonics functions. This expansion is exact and valid everywhere outside the smallest sphere enclosing all charges. For a collection of n point charges q_i at positions $\mathbf{x}_i$, the coefficients M_{lm} are given by

$$M_{lm} = \sum_{i=1}^{n} q_i Y_{lm}^*(\theta_i, \phi_i) r_i{}^l, \tag{12}$$

where Y_{lm}^* denotes the complex conjugates of the spherical harmonics functions. In addition to the multipole expansion in equation 11, the FMA also uses the local expansion

$$\Psi(\mathbf{x}) = 4\pi \sum_{l,m} L_{lm} Y_{lm}(\theta, \phi) r^l, \tag{13}$$

which is equivalent to a Taylor expansion and valid within the largest sphere around the origin that does not contain any of the charges contributing to it.

The expansions in equations 11 and 13 are exact, i.e. within their regions of validity they provide the exact value of the potential. However, they also consist of an infinite number of terms. The approximation in the FMA consists of truncating the expansions after a certain number of terms with the truncation limit p being the largest value of l included in the sums. It has been shown that the same level of truncation is appropriate for both the multipole and the local expansions. The essential idea of the FMA is to build a hierarchy of multipoles, each containing the contribution of a subset of charges of limited extent. For efficiency, cubic boxes are used in the FMA. The smallest boxes on the lowest level contain only a small number of charges (10–20), the boxes on successively higher levels are the union of eight lower level boxes, until one single box on the highest level contains all other boxes, and therefore all charges. The number of levels L is chosen to provide the optimal tradeoff between multipole and direct interactions and depends on the number of charges in the system as $N \sim 8^L$.

The multipoles associated with the lowest level boxes are calculated according to equation 12. Multipoles of higher level boxes are calculated not from the charges, but from the multipoles of the lower levels. To obtain the multipole expansion for one box, the eight multipole expansions of its subboxes are translated to a common origin and added up. Translating the origin of a multipole expansion by a translation vector $\mathbf{x}_t$ involves finding the coefficients of $\Phi'(\mathbf{x})$ such that

$$\Phi'(\mathbf{x}) = \Phi(\mathbf{x} + \mathbf{x}_t). \tag{14}$$

The relationship between the old and new coefficients is linear, i.e.

$$M'_{lm} = \sum_{l',m'} T^{MM}_{lm,l'm'} M_{l'm'}, \tag{15}$$

and the translation matrix is given by

$$T^{MM}_{lm,l'm'} = 4\pi \frac{(2l+1)\, a_{l'm'} a_{l-l',m-m'}}{(2l'+1)\,(2l-2l'+1)\, a_{lm}} \, (-r_t)^{l-1} Y^*_{l-l',m-m'}(\theta_t, \phi_t), \tag{16}$$

with the auxiliary numbers a_{lm} defined as

$$a_{lm} = (-1)^{l+m} \sqrt{\frac{2l+1}{4\pi\,(l+m)!\,(l-m)!}} \tag{17}$$

Since the coefficients $T^{MM}_{lm,l'm'}$ depend solely on the translation vector $\mathbf{x}_t$, and since in the regular cubic arrangement of multipoles the same translation vectors occur many times, the coefficients can be precomputed for the L levels and 8 possible directions and then reused very efficiently during the calculations.

After the multipoles $\Phi(\mathbf{x})$ of all boxes have been calculated, a local expansion $\Psi(\mathbf{x})$ is constructed for each box, describing the potential inside the box caused by all distant charges, i.e. all charges except those in the same or neighboring boxes. An essential element of the algorithm is the recursive use of the local expansion from the next higher level, i.e.

$$\Psi_l(\mathbf{x}) = \Psi'_l(\mathbf{x}) + \Psi_{l-1}(\mathbf{x}), \tag{18}$$

where $\Psi_l'(\mathbf{x})$ is the local expansion of all multipoles not already contained in $\Psi_{l-1}(\mathbf{x})$. The number of those multipoles is never larger than $6^3 - 3^3 = 189$, thus the number of operations for this step is independent of the size of the system for each box. Since the number of boxes depends linearly on the number of charges, this step, like all others, is scalable. The coefficients of the local expansion of a given multipole expansion are given by

$$L_{lm}' = \sum_{l',m'} T_{lm,l'm'}^{LM} M_{l'm'} \tag{19}$$

with the transformation matrix

$$T_{lm,l'm'}^{LM} = 4\pi \frac{a_{l'm'} a_{lm}}{(2l'+1)(2l+1) a_{l+l',m-m'}} \frac{(-1)^{l'+m'}}{r_t^{l'+l+1}} Y_{l+l',m-m'}^*(\theta_t, \phi_t), \tag{20}$$

and they can be precomputed in the same way as $T_{lm,l'm'}^{MM}$. Most of the time spent evaluating the multipole interactions is in this transformation. The translation of $\Psi_{l-1}(\mathbf{x})$ to the origin of $\Psi_l(\mathbf{x})$ is done similar to the translation of multipoles, i.e.

$$L_{lm}' = \sum_{l',m'} T_{lm,l'm'}^{LL} L_{l'm'}. \tag{21}$$

with the transformation matrix

$$T_{lm,l'm'}^{LL} = 4\pi \frac{a_{lm} a_{l'-l,m'-m}}{(2l+1)(2l'-2l+1) a_{l'm'}} (-r_t)^{l'-l} Y_{l'-l,m'-m}(\theta_t, \phi_t). \tag{22}$$

The local expansions $\Psi_l(\mathbf{x})$ are calculated in a hierarchical, sequential manner such that only one set of expansion coefficients need to be stored per level at any one time, thereby greatly reducing the storage requirement for the FMA. Once the local expansions at the lowest level $\Psi_L(\mathbf{x})$ are known, the interaction energy of a particle $\mathbf{x}_i$ with all other atoms not within neighboring boxes is given as

$$U_\psi(\mathbf{x}_i) = \frac{q_i}{\epsilon} \Psi_{L,i}(\mathbf{x}_i), \tag{23}$$

where $\Psi_{L,i}$ is the local expansion in the box of particle i. The total Coulomb potential is then

$$U_{\text{Coulomb}} = \sum_{i=1}^{N} \frac{q_i}{\epsilon} \Psi_{L,i}(\mathbf{x}_i) + \sum_{\langle ij \rangle} \frac{q_i q_j}{\epsilon r_{ij}}. \tag{24}$$

The pair sum $\langle ij \rangle$ in the second term is restricted to pairs within one box or between neighboring boxes.

The parallel implementation of the FMA is currently quite straightforward and not entirely scalable in storage and computation. The atom coordinates are distributed globally during the long range step, and the FMA is divided into a number of tasks, which are assigned to processing nodes according to balancing requirements. The tasks are created by restricting the FMA to subvolumes of the total grid and calculating the local expansions only in that subvolume. All multipoles are calculated on each node. This simple technique eliminates the need for communication of multipoles, but it is limited to about 1 million atoms and a few hundred processors. A fully scalable parallel version of the FMA is under development and will relieve these limitations by making use of data distribution and a hierarchical multipole communication scheme.

The Distance Class Algorithm

The Fast Multipole Algorithm makes the calculation of the full Coulomb interactions linear in complexity, but it still takes considerably more time to execute than a cutoff based calculation. The Distance Class Algorithm (DCA) has been developed for PMD to reduce the computation time per timestep further such that it is comparable in cost with the much less accurate cutoff methods. To keep track of the short range interactions, a pairlist is maintained and updated regularly using a very efficient scalable, grid based algorithm. The DCA differs from other multiple timestep methods (10, 11, 18, 19) mainly in that it is simpler. Application of more advanced methods could provide better accuracy and is being considered for future versions of PMD.

In molecular dynamics simulations, fast motions are generally vibrations or rotations of atoms that don't take the atoms further than about 1Å from their average position. This is a property of the force field that is due to the restrictive nature of bonded interactions and van der Waals interactions which prohibits the atoms from moving freely. The average positions around which atoms vibrate change much more slowly. Because of this separation of time scales, the Coulomb interaction between distant atoms changes very little during a short period of time. It is thus a good approximation to separate interactions into distance classes and to keep the potential and forces resulting from all but the first class constant for several timesteps.

The Coulomb interactions are divided into N_c different terms

$$U_{\text{Coulomb}} = \sum_{k=1}^{N_c} \sum_{\{i,j\}_k} \frac{q_i q_j}{\epsilon r_{ij}} \tag{25}$$

by decomposing the set of all pairs $\{i,j\}$ into N_c disjoint sets $\{i,j\}_k$ called *distance classes*. The decomposition is performed by first specifying a number of distance values $d_1, d_2, \ldots d_{N_c}$ such that

$$d_1 < d_2 < \ldots < d_{N_c} = \infty \tag{26}$$

and then classifying all pairs of atoms into classes according to their distance r_{ij} in space

$$\{i,j\}_k = \{(i,j) : d_k < r_{ij} \le d_{k+1}\}. \tag{27}$$

The Distance Class Algorithm in PMD uses two classes, one for short range interactions and one for long range interactions. The short range interactions are calculated from the same list of atom pairs that is used to calculate the van der Waals interactions, which are truncated at the cutoff distance d_c. The long range interactions are calculated by subtracting the short range potential and forces from the full potential and forces calculated using the Fast Multipole Algorithm. They are then stored and applied as a constant force until either an atom is detected to move more than a given tolerance d_{tol}, or until a certain number of steps N_{int} is counted since the last update. The interval of constant force N_{int} need not be longer than about 20, at which point the FMA becomes comparable in CPU time usage to the short range interactions. The tolerance d_{tol} ensures that there is an upper bound on the error incurred by the DCA.

This combination of FMA and DCA makes it possible to calculate the full Coulomb interactions of large systems in a time comparable with conventional cutoff calculations. Indeed, since the long range interactions are not neglected, the class separation distance can be chosen shorter than the typically used cutoff distance, making the FMA/DCA method both more accurate and faster than cutoff calculations. Fig. 2 illustrates the performance gain compared to a direct pairwise full Coulomb calculation.

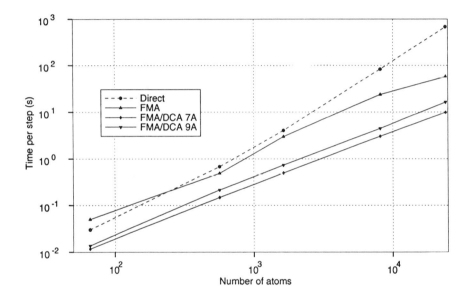

Figure 2: Performance of the FMA and the combined FMA/DCA methods compared to a full pairwise Coulomb calculation, plotted logarithmically against system size. Curves for $d_c = 7$Å and $d_c = 9$Å are shown. With 24,000 atoms, almost a factor of 100 is gained by using FMA and DCA.

The tolerance is chosen to be $d_{tol} = 1$Å, which provides on the order of 20 steps in which the pairlist does not need to be rebuilt. In accordance to that, and because larger intervals do not gain a significant performance increase, the maximum pairlist rebuild interval is set to $N_{int} = 20$. This provides an accuracy better than 1% for the electrostatic forces compared with full calculation at each step. This accuracy corresponds well with the accuracy of the FMA truncated at $p = 4$ terms, which is about 0.5%.

The evaluation of short range interaction, i.e. all interactions between atoms no more than the cutoff distance d_c apart, is done most efficiently using a pairlist. For each atom a list is constructed that contains the numbers of all atoms within the cutoff distance. In order for the procedure to be scalable, a grid algorithm is used to preselect candidates for the pairlist, cutting the time for constructing the pairlist to $O(N)$. Since each node needs only the pairlist for local atoms, the parallel scaling of pairlist generation is $O(N/P)$, satisfying the scalability requirement. When the pairlist is updated, the distance classes change and the long range interactions have to be recalculated. It is therefore natural and necessary to couple pairlist generation and long range force evaluation to always occur together.

For efficient pairlist generation, a cubic grid is constructed, with a grid constant equal to the cutoff distance. Each atom is assigned to the closest grid point, producing a list of atoms at every grid point. Next, a loop through neighboring pairs of grid points is executed and for each pair of grid points all possible atom pairings are examined. Those atom pairs that are within the cutoff distance are entered into the pairlist. Only half of the neighbors of a grid point need to be considered, since each pair of atoms would otherwise be listed twice.

In the parallel case, communication time can be reduced by adding an additional criterion to the pair selection. Each pair of neighboring processing nodes has a directed link assigned to it. This link points to the node which will calculate all non-bond interactions across the interface between the node regions. Since a node needs to know only the coordinates of those ghost atoms with which it is to calculate the interactions, communication can be cut in half by not updating unused ghost atom coordinates. Thus, the rules for including pairs in the pairlist are as follows: For local pairs, an atom is paired with another only if its grid point number is higher. For non-local pairs, an atom is paired with another only if its node is responsible for calculating the interaction. The first criterion allows the loop over grid points to be cut in half, the second reduces the ghost atom communication.

While the additional criteria complicate the code, they have no effect on the efficiency of the algorithm when only one node is used, and they greatly reduce communication when the code is run in parallel.

Solvent accessible area and the Circle Intersection Method

Simulation of macromolecules in vacuum is rarely an appropriate model for the real system. Solvent effects are crucial for determining protein structure and for almost all biological processes (20–24). Solvent effect can be roughly divided into two areas: Hydrophobic effect and electrostatic properties. The hydrophobic effect is due to the self interaction of water, and occurs when water would rather be next to itself than next to a solute molecule. The electrostatic effect derives from the high dielectric constant of water, which attenuates and attracts electric fields, leading to suppressed Coulomb interactions and dielectric pressure forces. There is also an electrostatic effect from ions in the solution, which further attenuates electric fields.

The easiest way to model solvent is explicit solvent simulation. A large number of water molecules are included in the system, and given an appropriate water model, all properties of the solvent-solute interactions should emerge from the microscopic model. While this is feasible, it is also very expensive in terms of computer time. PMD is well suited for explicit solvent calculations, since its scalability provides for the large and very large systems that generally result from building explicit solvent models. However, it is often desirable to use more sophisticated models, and the future development of PMD is directed towards providing both implicit solvent models as well as continuum electrostatics methods for solvent modeling.

Implicit solvent models attempt to reproduce solvent effects by defining a potential, often with a number of parameters fitted to observations, that can be easily calculated from atom positions in the same manner as the other non-bonded interactions. Several such models have been proposed, some based on solvent accessible surface areas (25–29), others based on atom coordinates only (30–32).

Continuum electrostatics is a more rigorous aproach to calculating the electrostatic part of the solvent effects (24). It is based on solving the Poisson Boltzman equation with the solute represented as a low-dielectric cavity in a high dielectric medium, with the molecular surface separating the two regions. Most commonly a finite difference method is used to solve the Poisson Boltzman equation (20,33–35), but the boundary element method is also used (36–40).

All continuum electrostatics methods as well as most implicit solvent methods depend on a representation of the solvent accessible or molecular surface, which separates the interior solute volume from the exterior solvent space. Thus the first step towards solvent modelling should be a fast and scalable method for the computation of molecular surface areas and their derivatives. PMD currently implements the *Circle Intersection Method* (CIM), a fast method for the calculation of the solvent accessible surface of the solute, including the derivatives to obtain forces. The CIM essentially follows Conolly (41) with regards to the geometry of the surface, but uses a novel method for finding vertices. The computation time required is 2 3 milliseconds per atom on a SGI Indigo 2/R4400 and scales linearly with the number of atoms for arbitrarily large molecules.

The CIM is estimated to be about thirty times faster than the original method ANA, and approaches the MSEED method in speed (42). Unlike MSEED, however, PMD measures the complete surface including cavities with correct treatment of complete circles of intersection, which have been found to occur quite frequently in larger proteins. Table 1 shows a comparison of PMD with MSEED and ANAREA, a method developed by Richmond (43) and modified by Wesson (27). ANAREA was the fastest exact analytic program available for this comparison. The Brookhaven Protein databank files for crambin (1crn, 327 atoms), pancreatic trypsin inhibitor (1pti) with added hydrogens (568 atoms), T4 phage lysozyme (2lzm, 1427 atoms), MHC class I receptor (1vaa, 3235 atoms), a poliovirus shell protomer (2plv, 7162 atoms), and the photosynthetic reaction center (1prc, 10288 atoms) were used. In those cases where full circle intersections don't exist (1crn, 1pti, and 2lzm), the PMD result for the outer surface area agrees almost perfectly with MSEED. Where full circles exist, MSEED erroneously reports larger surface areas. PMD and ANAREA both measure the surface including cavities, and agree well for all test cases. The agreement is not as good as the one between PMD and MSEED, indicating that ANAREA is somewhat less accurate than PMD and MSEED. The CIM and MSEED are both scalable, i.e. scale as $O(N)$, while ANAREA scales as $O(N^2)$ and will slow down more and more as molecules get larger.

Ultimately, PMD is planned to provide solvation modelling with the Finite Difference Poisson Boltzmann (FDPB) method for the electrostatics and the CIM surface for the hydrophobic effect. Several methods have been suggested for de-

Table 1: Solvent accessible areas of proteins calculated with three different programs. Areas were calculated with PMD, with ANAREA, with MSEED, and with PMD omitting the cavities (pmd–cav). The ANAREA code used was not dimensioned large enough for the photosynthetic reaction center (10288 atoms).

	Accessible surface area				CPU time (ms/atom)		
Atoms	pmd	anarea	mseed	pmd–cav	pmd	anarea	mseed
327	2976.4604	2976.4604	2974.2777	2974.2777	2.23	5.05	1.65
568	4031.4016	4031.4011	4021.3169	4021.3169	2.22	5.66	1.42
1427	9123.6446	9123.6455	9098.2890	9098.2891	2.40	6.22	1.79
3235	17975.5711	17975.5840	17903.8977	17901.2460	2.58	6.73	2.04
7162	35036.5994	35036.7813	34854.0903	34808.7272	2.77	7.64	1.77
10288	43824.2222	n/a	43313.6653	43294.5298	3.07	n/a	1.97

riving forces from continuum electrostatics methods (44–46), demonstrating the feasibility of that approach.

Implementation and Performance

PMD was designed to be modular and very portable. The structure of the program is object oriented, although the programming language used is ANSI C, in the interest of portability. Fig. 3 shows an overview of the basic parts of PMD. Currently, PMD is most useful as a library of subroutines, a high level user interface is not yet included.

All aspects of data distribution and parallelism are contained in the Data Distribution and Parallel Adaptor modules. The parallel adaptors, particularly, define a small, simple set of routines that are used exclusively to access the message passing features of the underlying parallel architecture. This makes PMD extremely portable from one parallel machine to another. Reimplementation of the parallel adaptor can be done in a few hours.

The modular design of PMD provides a general framework to which new functionality can be easily added without need to change most of the program. Individual modules can be replaced by completely different implementations without altering the function of the rest of the system. For instance, in the course of development the Data Distribution module was completely redesigned to introduce load balancing, switching from a cubic grid distribution to the current distribution based on Voronoi polyhedra. None of the modules built on the Data Distribution module needed to be changed significantly and they continued working with the new distribution.

Central to this modularity is the definition of an abstract interface defining how application modules are to access the underlying PMD data structures, without having to be aware of the form of data distribution, or the fact that data is distributed at all. This interface has evolved during the development of PMD, and documentation will be forthcoming when the specifications have sufficiently solidified. This will allow developers to use PMD as a library from which to develop their own molecular modelling applications without having to know the PMD code in detail, and to take advantage of the infrastructure for parallelism and molecular structure description of PMD without undue extra effort.

PMD is continually under development, and source code versions are made available periodically to encourage researchers to use PMD for their own purposes. The only conditions attached to the distribution are 1) that no part of it be used in commercial applications without prior consent of the author, 2) that any additions or modifications be made freely available on the same basis as the original code. The current version of the PMD distribution can be obtained by anonymous ftp from the machine *cumbnd.bioc.columbia.edu*. An electronic mailing list for announcements of interest to users of PMD is maintained at Columbia University. To subscribe, send electronic mail to *pmd-request@cumbnd.bioc.columbia.edu*, or contact the author of this paper.

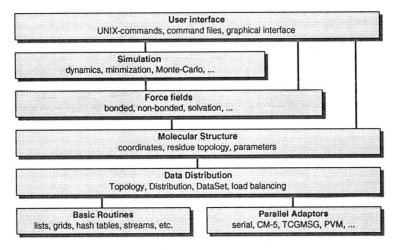

Figure 3: Diagram of the principal components of PMD. Not everything mentioned in the graph is actually implemented, particularly there is no graphical interface, no PVM adaptor (which is trivial to add) and no solvation force field.

PMD was implemented and tested on a variety of workstations, and parallel adaptors exist for workstation networks, for the Thinking Machines CM-5 and for the Intel Paragon. The network and the Paragon implementation both use a TCGMSG adaptor, the CM-5 has a specific adaptor. Implementations for PVM and the Cray T3D are planned. Donations of development time on any kind of message passing parallel machine will be gratefully accepted and quickly rewarded with a PMD implementation for that machine.

To evaluate the performance of PMD both on parallel machines and on common workstations, test calculations were done using a realistic model of a POPC lipid bilayer patch with water consisting of 23,975 atoms. Some results are reported in table 2. No results are given for the CM-5, since PMD does not use the vector processors of that machine, rendering the benchmark meaningless. However, PMD was run successfully, if not fast, with 256 processors on the CM-5, demonstrating the high degree of parallelism possible with PMD.

The communication code associated with the Voronoi decomposition topology is not well optimized yet, but, nevertheless, quite respectable speedups can already be achieved on the Intel Paragon, as well as on a network of HP-735 workstations interconnected using an ATM switch. Fig. 4 shows the dependence of the calcula-

Table 2: Benchmark results for a realistic model of a POPC lipid bilayer patch consisting of 23,975 atoms. Times are given in seconds for the FMA calculation, the pairlist update and the evaluation of short range electrostatic and van der Waals interactions. Also shown is the average total time per step, assuming a long range update interval of $N_{int} = 20$. The ALR is a Pentium desktop PC running NEXTSTEP, the HP-735 ATM entries refer to a workstation cluster connected via an ATM switch, using 4 resp. 8 machines.

System	PFMA	pairlist	short	average
ALR 586/60 NeXT	150	40	18	27.50
SUN SPARC-10	110	30	14.5	21.50
SGI Indigo/R4000	90	21	13	18.55
DEC 3000/500	75	14	10	14.45
IBM-590	55	27	7.4	11.50
SGI Indigo/R4400	55	15	7.6	11.10
SGI Onyx/150MHz	54	14	7.3	10.70
HP 712/80MHz	57	19	6.8	10.60
HP 735/100MHz	34	11.5	3.9	6.18
Intel Paragon(16)	20	3.2	2.2	3.36
HP-735 ATM(4)	10	2.7	1.1	1.74
HP-735 ATM(8)				1.10

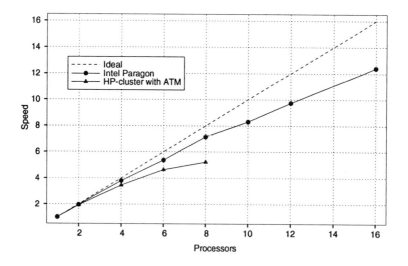

Figure 4: Parallel benchmark results for the Intel Paragon parallel computer and for a cluster of HP-735 workstations connected via an ATM switch. The same 23,975 atom system is used as in table 2. The difference in scaling between the workstation cluster and the Paragon is due primarily to the difference in speed between the individual nodes rather than to the communications network. These results are based on new, non-optimized communication code, scaling is expected to improve in future releases.

tion time on the number of nodes. At least 50% of the communication can still be eliminated by a more optimized implementation which will hopefully be available in the next release of PMD.

Discussion and Outlook

It should be clear from this presentation that PMD is not a finished program, but should be viewed as a workbench for algorithm development in molecular modelling. The features that make it unique are that it is fast, fully scalable and does not neglect long range interactions. Using the FMA, PMD removes the tradeoff that has traditionally been required between speed and long range interactions, and by its scalable nature it also removes most barriers to the simulation of very large systems. Moreover, PMD makes large parallel supercomputers more accessible, since the same program that runs on a desktop workstation will run identically on the parallel machine.

Since the basic algorithms and data distribution mechanisms are largely in place, future developments will concentrate on application oriented issues. Mechanisms are needed to build structures of proteins and explicit solvent models when coordinates are not known. Algorithms to add hydrogens and sidechains to protein backbones are being considered. Mostly, however, PMD will be extended to provide better solvation treatment, on the basis of solvent accessible surface areas and the Poisson Boltzmann equation. Methods for the derivation of forces from continuum electrostatics will be explored.

Acknowledgment

This work was supported by the National Center for Research Resources division of the Biomedical Technology Program at the NIH, through a Research Resource grant (P41 RR06892) at Columbia University and NSF grant (DIR-9207256) to Barry Honig. An early prototype of PMD was developed with support by Klaus Schulten of the NIH Resource for Concurrent Biological Computing at the University of Illinois at Urbana-Champaign, who also provided computer time for the ATM cluster benchmarks. Tim Mattson of Intel Corp. generously provided computer time and advice on the Intel Paragon parallel supercomputer.

Literature Cited

1. Levitt, M.; Lifson, S. *J. Molec. Biol.* **1969**, *46*, 269.
2. Karplus, M.; McCammon, J. A. *Ann. Rev. Biochem* **1983**, *53*, 263.
3. Karplus, M. Molecular dynamics of proteins. In *Structure and Dynamics of Nucleic Acids, Proteins, and Membranes*; Clementi, E.; Chin, S., Eds., pp 113–126, London, 1986. Plenum Press.
4. Clementi, E.; Corongiu, G.; Aida, M.; Niesar, U.; Kneller, G. In *MOTECC, Modern Techniques in Computational Chemistry*; Clementi, E., Ed., pp 805–882. ESCOM, Leiden, 1990.
5. Weiner, P. K.; Kollman, P. A. *J. Comp. Chem.* **1981**, *2*, 287–303.
6. Brooks, B. R.; Bruccoleri, R. E.; Olafson, B. D.; States, D. J.; Swaminathan, S.; Karplus, M. *J. Comp. Chem.* **1983**, *4*(2), 187–217.

7. van Gunsteren, W. F. *GROMOS:Groningen Molecular Simulation Program Package.* University of Groningen, The Netherlands, 1987.

8. Loncharich, R. J.; Brooks, B. R. *Proteins* **1989**, *6*, 32–45.

9. Board, Jr., J. A.; Causey, J. W.; Leathrum, Jr., J. F.; Windemuth, A.; Schulten, K. *Chem. Phys. Lett.* **1992**, *198*, 89–94.

10. Grubmüller, H.; Heller, H.; Windemuth, A.; Schulten, K. *Molecular Simulation* **1991**, *6*(1–3), 121–142.

11. Teleman, O.; Jönssen, B. *J. Comp. Chem.* **1986**, *7*, 58–66.

12. Plimpton, S.; Hendrickson, B. Technical Report SAND94–1862, Sandia National Laboratories, 1994.

13. Plimpton, S. Technical Report SAND91–1144, Sandia National Laboratories, 1991.

14. Greengard, L.; Rohklin, V. *J. Comp. Phys.* **1987**, *73*, 325–348.

15. Greengard, L. *The Rapid Evaluation of Potential Fields in Particle Systems.* MIT Press, Cambridge, MA, 1988.

16. Leathrum, Jr., J. F.; Board, Jr., J. A. Technical report, Duke University, Department of Engineering, 1992.

17. Ding, H.-Q.; Karasawa, N.; Goddard, W. A. *J. Chem. Phys.* **1992**, *97*, 4309–4315.

18. Tuckerman, M. E.; Berne, B. J.; Martyna, G. J. *J. Chem. Phys.* **1991**, *94*, 6811–6815.

19. Scully, J. L.; Hermans, J. *Molecular Simulation* **1993**, *11*, 67–77.

20. Perutz, M. F. *Science* **1978**, *201*, 1187–1191.

21. Warshel, A.; Russell, S. T. *Q. Rev. Biophys.* **1984**, *17*, 283.

22. Honig, B.; Hubbell, W.; Flewelling, R. *Ann. Rev. Biophys. Biophys. Chem.* **1986**, *15*, 163–193.

23. Jayaram, B.; Sharp, K. A.; Honig, B. *Biopolymers* **1989**, *28*, 975–993.

24. Sharp, K. A.; Honig, B. *Ann. Rev. Biophys. Biophys. Chem.* **1990**, *19*, 301–332.

25. Eisenberg, D.; McLachlan, A. D. *Nature* **1986**, *319*, 199–203.

26. Eisenberg, D.; Wesson, M.; Yamashita, M. *Chem. Scripta* **1989**, *29A*, 217–221.

27. Wesson, L.; Eisenberg, D. *Protein Science* **1992**, *1*, 227–235.

28. Schiffer, C. A.; Caldwell, J. W.; Stroud, R. M.; Kollman, P. A. *Protein Science* **1992**, *1*, 396–400.

29. von Freyberg, B.; Richmand, T. J.; Braun, W. *J. Molec. Biol.* **1993**, *233*, 275–292.

30. Still, W. C.; Tempczyk, A.; Hawley, R. C.; Hendrickson, T. *J. Am. Chem. Soc.* **1990**, *112*, 6127–6129.

31. Stouten, P. F. W.; Frömmel, C.; Nakamura, H.; Sander, C. *Molecular Simulation* **1993**, *10*, 97–120.

32. Davis, M. E. *J. Chem. Phys.* **1994**, *100*, 5149–5159.

33. Gilson, M. K.; Rashin, A.; Fine, R.; Honig, B. *J. Molec. Biol.* **1985**, *183*, 503–516.

34. Gilson, M. K.; Sharp, K. A.; Honig, B. *J. Comp. Chem.* **1987**, *9*, 327–335.

35. Rashin, A. A. *Int. J. Quantum Chem.: Quantum Biol. Symp.* **1988**, *15*, 103–118.
36. Zauhar, R. J.; Morgan, R. S. *J. Molec. Biol.* **1985**, *186*, 815–820.
37. Zauhar, R. J.; Morgan, R. S. *J. Comp. Chem.* **1988**, *9*, 171–187.
38. Zauhar, R. J.; Morgan, R. S. *J. Comp. Chem.* **1990**, *11*, 603–622.
39. Rashin, A. A. *J. Phys. Chem.* **1988**, *94*, 1725–1733.
40. Vorobjev, Y. N.; Grant, J. A.; Scheraga, H. A. *J. Am. Chem. Soc.* **1992**, *114*, 3189–3196.
41. Conolly, M. L. *J. Appl. Cryst.* **1983**, *16*, 548–558.
42. Perrot, G.; Cheng, B.; Gibson, K. D.; Vila, J.; Palmer, K. A.; Nayeem, A.; Maigret, B.; Scheraga, H. A. *J. Comp. Chem.* **1992**, *13*, 1–11.
43. Richmond, T. *J. Molec. Biol.* **1984**, *178*, 63–89.
44. Gilson, M. K.; Honig, B. *Journal of Computer-Aided Molecular Design* **1991**, *5*, 5–20.
45. Zauhar, R. J. *J. Comp. Chem.* **1991**, *12*, 575–583.
46. Gilson, M. K.; Davis, M. E.; Luty, B. A.; McCammon, J. A. *J. Phys. Chem.* **1993**, *97*, 3591–3600.

RECEIVED November 15, 1994

Chapter 12

Parallelization of Poisson–Boltzmann and Brownian Dynamics Calculations

Andrew Ilin[1], Babak Bagheri[1], L. Ridgway Scott[1], James M. Briggs[2], and J. Andrew McCammon[2]

[1]Department of Mathematics and [2]Department of Chemistry, University of Houston, Houston, TX 77204

Poisson-Boltzmann calculations are increasingly used in chemistry and biochemistry to determine the electrostatic free energy of solute molecules in electrolyte solutions. The forces acting on such molecules can also be calculated and used in Brownian dynamics simulations of diffusional motion of the solutes. All of these calculations become computationally intensive as the model systems are described in greater detail. Here we describe recent advances in the parallelization of such calculations. Illustrative results are presented for the enzyme acetylcholinesterase.

Electrostatic interactions play a key role in determining the stability of conformations and complexes of solute molecules in solution. Because these interactions are long-ranged, they also play an important role in determining the rates of diffusional conformational change and of diffusional encounter of solute molecules.

The most accurate computational models of the systems listed above would include an atomistic description of both the solvent and solute molecules of interest. But such fully microscopic models, as employed for example in molecular dynamics simulations, can not yet be used in studies of many processes of interest because of current limitations in the performance and capacity of computers.

Fortunately, the solvent and secondary solute species such as spectator ions can be replaced to a good approximation in many cases by continuum models, and the continuum treatment can even be employed for the interiors of the solute species of primary interest (1). For example, simulations based on such models were recently used successfully to guide the first quantitative engineering of a faster enzyme that was then proven in the laboratory (2,3).

Simulations based on continuum-type models can still be very demanding of computational resources, however. For example, it is not yet possible to use the standard continuum method for calculating electrostatic forces (based on the

0097–6156/95/0592–0170$12.00/0
© 1995 American Chemical Society

Poisson-Boltzmann equation, see below) to recompute these forces during a Brownian dynamics simulation of the encounter of a substrate with an enzyme molecule. Instead, the substrate is typically treated as one or more test charges moving in the fixed field of the isolated enzyme; this approximation ignores certain interactions, such as those between the charges in the substrate and their image charges due to the dielectric interface between the enzyme and the solvent.

In the present paper, we outline progress in the development of algorithms and codes to speed continuum-type calculations on parallel computers. The computer software used is the University of Houston Brownian Dynamics (UHBD) program (1) and the computations are done on the Intel Delta machine. Here, we first describe in some detail how the electrostatic calculations can be parallelized efficiently. We then illustrate how the Brownian dynamics trajectory calculations can be parallelized, particularly by improvements in the generation of random numbers. We then illustrate the use of such methods for a system of great biomedical importance, the enzyme acetylcholinesterase.

Mathematical Model of Electrostatics

The Nonlinear Poisson-Boltzmann equation (NLPBE) can be used to calculate the electrostatic potential field $\phi(r)$ of a molecule (4, 5). The NLPBE can be written in the following dimensionless form for a univalent electrolyte solution:

$$-\nabla \cdot \varepsilon(r)\nabla\phi + k(r)\sinh\phi = \rho(r) \ in \ \Omega \in \mathbf{R}^3, \tag{1}$$
$$\phi(r) = \phi_0(r) \ on \ \partial\Omega,$$

where $\varepsilon(r)$ is the dielectric constant, $\rho(r)$ is the charge density, $k(r)$ is the dimensionless solvent ionic strength, r is a position, and $\phi_0(r)$ is assumed to be known on the boundary of the domain Ω. Typically $\varepsilon(r)$ and $k(r)$ are piecewise constant functions, $\rho(r)$ is a sum of Dirac δ-type singular functions, but $\phi(r)$ and $\varepsilon(r)\nabla\phi(r)$ are continuous.

Numerical method

When $k(r)$ has relatively small values, i.e. the NLPBE is not "too nonlinear" a conjugate gradient method based on a variational formulation of the given nonlinear problem can be used for solving Equation 1 (4).

The full nonlinear Poisson-Boltzmann equation with the given boundary conditions can be solved by a damped version of Newton's method:

$$\phi_{new} = \phi + \omega\xi \tag{2}$$

where the damping factor $\omega \in [0, 2]$ and the correction ξ is the solution of the linearized equation:

$$-\nabla \cdot \varepsilon(r)\nabla\xi + k(r)\cosh\phi(r)\xi = \rho(r) + \nabla \cdot \varepsilon(r)\nabla\phi(r) - k(r)\sinh\phi(r). \tag{3}$$

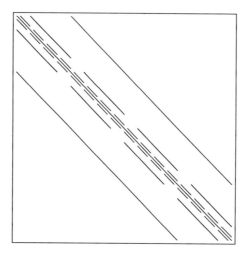

Figure 1: Sparsity of A

The convergence criterion for Newton's method is

$$\|\xi\| \leq \delta_N \|\phi\|,$$

where $\|\xi\|$ is some norm of correction ξ and δ_N is a convergence tolerance.

Typical difference methods for the solution of equation 3 lead to the algebraic system of equations

$$Ax = b, \tag{4}$$

where x is a discretization of the correction potential ξ, b is a discretization of the residual term in Equation 3 $(\rho + \nabla \cdot \varepsilon \nabla \phi - k(r) \sinh \phi)$, A is the sparse matrix corresponding to an approximation of the linear differential operator $-\nabla \cdot \varepsilon \nabla + k(r) \cosh \phi$.

In three dimensions and for the standard seven-point difference stencil, the matrix A consists of the following diagonals (6)

$$A = E + C + B + D + B^T + C^T + E^T = L + D + L^T. \tag{5}$$

shown in Figure 1, where L is the lower triangular matrix with three diagonals.

Numerical experiments have shown that the Preconditioned Conjugate Gradients (PCG) method, shown in Figure 2 (where M is the preconditioning matrix), is preferable to other iterative techniques (6–8) for solving this system of equations. We will consider diagonal and Incomplete Cholesky preconditioner M, having a sparsity (nonzero) structure subordinate to that of A. We can exploit parallelism in the following steps of one iteration of the PCG algorithm (which takes a total of $35N$ flops):

1. vector addition and multiplication by a constant ($6N$ flops),
2. vector multiplication $(r_i \cdot M^{-1} r_i)$ and $(p_i \cdot A p_i)$ ($4N$ flops),

$$r_0 = b - Ax$$
$$p_0 = M^{-1}r_0$$
For $i = 0$ to convergence
$$\alpha = (r_i \cdot M^{-1}r_i)/(p_i \cdot Ap_i)$$
$$x_{i+1} = x_i + \alpha p_i$$
$$r_{i+1} = r_i - \alpha Ap_i$$
$$\beta = (r_{i+1} \cdot M^{-1}r_{i+1})/(r_i \cdot M^{-1}r_i)$$
$$p_{i+1} = M^{-1}r_{i+1} + \beta p_i$$
End For

Figure 2: Preconditioned Conjugate Gradients

3. evaluation of Ap_i (13N flops),

4. solution of $p_i = Mr_i$ (12N flops, if A was prescaled by the method which we will describe later),

where N is the number of unknowns.

As we will see in the following sections, the effective introduction of parallelism into last item is the most challenging aspect of parallelizing PCG. Parallelizing of other items were studied in Ref. (9).

Parallel preconditioning

M is a preconditioning matrix which should be easily invertible. The more closely M resembles A, the fewer iterations will be required, but the more difficult will be the problem of inverting M. A preconditioner which is easy to implement is the diagonal $M = D^{-1}$. Diagonal preconditioning will not, however, increase the rate of convergence of the conjugate gradient method dramatically.

A more effective preconditioner is the Incomplete Cholesky (IC) factorization of A (10,11):

$$M = (L + \Delta)\Delta^{-1}(\Delta + L^T), \tag{6}$$

where the lower triangular matrix $L + \Delta$ consists of four diagonals, shown in Figure 3.

The entries of the diagonal matrix Δ are given by

$$\delta_{ijk} = d_{ijk} - \frac{b_{i-1,jk}^2}{\delta_{i-1,jk}} - \frac{c_{i,j-1,k}^2}{\delta_{i,j-1,k}} - \frac{e_{ij,k-1}^2}{\delta_{ij,k-1}}, \tag{7}$$

where $d_{ijk}, b_{i-1,jk}, c_{i,j-1,k}, e_{ij,k-1}$ are entries of diagonal matrices D, B, C, E respectively. Since the preconditioning matrix is given as a product of two triangular factors (Equation 6), the preconditioning phase requires the solution of two triangular systems.

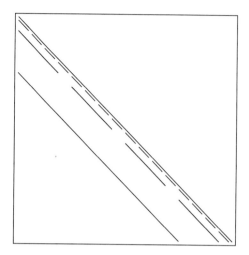

Figure 3: Sparsity of the lower triangular matrix of IC preconditioner

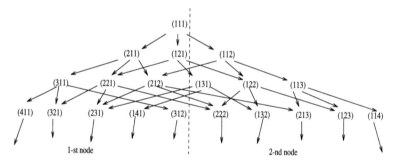

Figure 4: Graph of dependences of recurrence (8)

A straightforward way to parallelize triangular solves is to write the algorithm as a recurrence and schedule the computation to take advantage of whatever parallelism there is. For example, for the lower triangular factor, the recurrence has the following form:

$$p_{ijk} = r_{ijk} - b_{ijk}p_{i-1,jk} - c_{ijk}p_{i,j-1,k} - e_{ijk}p_{ij,k-1}. \tag{8}$$

The level-scheduled, multicolor and hyperplane methods (12–17) use the graph of dependences of the recurrence equation 8 (shown in Figure 4) to discover parallelism. For example, the level-scheduled method sorts the dependence graph topologically and distributes the calculation of each level among as many processors as possible. These approaches have the following disadvantages:

- considerable amount of communication between processors,
- bad load balance,

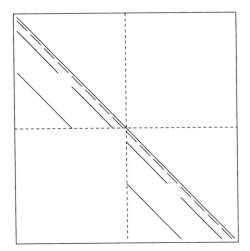

Figure 5: Sparsity of "parallelizable" IC preconditioner

- an order of computation different from the sequential algorithm.

A result of the first two disadvantages is that these methods are not scalable. Thus solving a 2-D problem on 4 processors gave a speedup factor of only 2.1 (16).

The approach we present here is to avoid the complete solution of the triangular systems. It is equivalent to introducing additional incompleteness into the IC preconditioner so that the matrix $L_p + \Delta$ is block-diagonal (see Figure 5). Notice that if the number of grid-points in the z-direction is a multiple of the number of processors then the "parallelizable" matrix L_p differs from the original one in only some part of the third subdiagonal.

The described parallel ICCG method (PICCG) has the following advantages:

- no communication between processors during the preconditioning phase,

- perfect load balance,

- no reordering of the computation.

The disadvantage of this method is the dependence of the convergence rate on the number of processors. We note that the parallel algorithm is not equivalent to the sequential algorithm.

Parallelization of the UHBD program for the Delta was performed using the IPfortran compiler (18). Our approach to the parallelization of the electrostatic phase is related to domain decomposition techniques. All domain data are distributed among processors. This requires data exchange between processors for evaluation of Ap. Also, vector multiplication requires global summation (reduction) of local results computed by each processor. The parallel ICCG method described above has also been used for the parallel mosfet simulation (19).

Scaling and Modification

To reduce the number of operations per iteration, we can scale A symmetrically by

$$A' = \Delta^{-1/2} A \Delta^{-1/2}, \tag{9}$$

so that the preconditioner M' of the scaled matrix A' has the identity matrix as its diagonal $\Delta' = I$. With this change the Equation 4 becomes

$$A'x' = b', \tag{10}$$

where $x' = \Delta^{1/2}x, b' = \Delta^{-1/2}b$.

To reduce the number of iterations, we can introduce a parameter α into the preconditioner by modifying its diagonal as follows.

$$
\begin{aligned}
\delta_{ijk} = d_{ijk} &- \frac{b_{i-1,jk}(b_{i-1,jk} + \alpha(c_{i-1,jk} + e_{i-1,jk}))}{\delta_{i-1,jk}} \\
&- \frac{c_{i,j-1,k}(c_{i,j-1,k} + \alpha(b_{i,j-1,k} + e_{i,j-1,k}))}{\delta_{i,j-1,k}} \\
&- \frac{e_{ij,k-1}(e_{ij,k-1} + \alpha(b_{ij,k-1} + c_{ij,k-1}))}{\delta_{ij,k-1}}
\end{aligned}
\tag{11}
$$

The resulting matrix is known as the Modified Incomplete Cholesky (MIC) preconditioner (7, 17). One should choose the value of α to minimize the number of iterations. Numerical experiments show that $\alpha = 0.95$ is often optimal for the one processor case and corresponding MICCG method converges two times faster than with ICCG. For a large number of processors the optimal α gets smaller and convergence of MICCG and ICCG is about the same (9).

Eisenstat's implementation

It has been shown that a significant reduction in CPU time is obtained for the symmetrically scaled (M)ICCG algorithm using Eisenstat's implementation (17). Introduced below is a similar improvement to our parallel (M)ICCG.

Let matrix $A = L + D + L^T$ be symmetrically prescaled according to Equation 9 and $M = (L_p + I)(I + L_p^T)$ be the corresponding parallel preconditioner. The idea of Eisenstat's implementation is to form the preconditioning matrix explicitly rather than using the preconditioner for redefining of the inner products for r_i as in Figure 2.

Consider the following system:

$$A'x' = b', \tag{12}$$

where $A' = (L_p + I)^{-1}A(I + L_p^T), x' = (I + L_p^T)x, b' = (L_p + I)^{-1}b$. One may check that $M^{-1}A = P^{-1}A'P$ with $P = I + L_p^T$ so A' and $M^{-1}A$ have the same condition numbers. This means that the preconditioned conjugate gradient method for Equation 4 has the same convergence rate as the pure conjugate gradient method for the Equation 12.

So we may omit the preconditioning phase and calculate $A'p'$ using the following formula:

$$A'p' = t' + (L_p + I)^{-1}((L - L_p + D - 2I + L^T - L_p^T)t' + p'), \qquad (13)$$

where $t' = (I + L_p^T)^{-1}p'$. Note that if the number of processors p is 1, then the computation of $A'p'$ costs $15N$ flops. For $p > 1$ it costs $17N$ flops; $8N$ flops for one conjugate gradient iteration is saved since original ICCG iteration costs $25N$ flops.

Brownian Dynamics

In a typical application, the phase of UHBD which computes Brownian Dynamics trajectories simulates the movement of a test particle toward the molecule being studied. The BD algorithm uses the electrostatic potential to compute trajectories of a diffusing test particle under the influence of a combination of random and electrostatic forces. A detailed description of the implemented algorithm is available (20).

The issues which make Brownian dynamics an interesting parallelization problem are

- deciding how to store the electrostatic potential needed to compute forces,
- balancing the computational load, and
- generating independent streams of random numbers.

Parallelization of Brownian dynamics is performed such that all trajectories are distributed among processors. Since each trajectory may travel through the whole domain, each processor should have access to electrostatic data in the entire domain. This means that before Brownian dynamics is started, the program should make the "local" electrostatic data "global". On the other hand, once the electrostatic simulation is over, the space which was occupied by working conjugate gradient arrays (like the residual vector r and others) can then be used for electrostatic potential storage. Limitations of memory on each processor dictated that the maximum mesh size is equal 100^3. This memory restriction may be removed if some type of distributed-shared memory (21) or paging memory were available.

Main Steps of Brownian Dynamics Simulation

Consider the method of determining the diffusion-controlled rate constant for the simple case of a symmetric two-particle system without electrostatic or hydrodynamic interactions. The Ermak-McCammon equation for the Brownian dynamics step of the particle reduces in this case to

$$r = r^0 + R + (k_B T)^{-1} DF(r^0)\Delta t, \qquad (14)$$

where r^0 is the relative position vector before a time step is taken, r is the relative position vector after a time step, and R is the vector of Gaussian random numbers of zero mean and variance

$$< R_i R_j >= 2D\delta_{ij}\Delta t, \ i = 1, 2, 3; \ j = 1, 2, 3. \tag{15}$$

D is the relative diffusion constant, Δt is the time step, k_B is Boltzmann's constant and T is the temperature.

A set of trajectories is simulated starting on the b-surface ($|r| = b$) for $b >> d$, (d is the diameter of each particle). Trajectories are terminated in the case of reaction ($|r| < d$) or in the case of particle escape ($|r| > q, q >> b$). Several hundred to several thousand trajectories are run to calculate a reaction probability β

$$\beta = \frac{N_r}{N}, \tag{16}$$

where N is the total number of trajectories and N_r is the number of reactive trajectories. The diffusion-controlled rate constant, reduced by the Smoulchowski result, is

$$k_r = \frac{2b\beta}{d(1 - (1 - \beta)\Omega)}, \tag{17}$$

where the quantity $\Omega = b/q$. For the simple two-particle case described above, the theoretical value of k_r is 1.0. Since a theoretical result is known, this is a convenient test of the random number generator.

Random Number Generator

When Brownian-dynamic trajectories are calculated in parallel, the streams of random numbers that are generated (22, 23) in parallel must have low correlation. If they do not, then the work done in parallel may be in vain. Very similar trajectories may have been calculated and hence resulting in little extra information. By simply adding the logical number of each process (or processor) to the seed used for random number generation, reasonably independent streams of random numbers are produced. This can be proven by computing the correlation function for two such streams explicitly.

Uniform random numbers can be used to generate a Gaussian random number. Currently the generator of G. Marsaglia and A. Zaman (22) is used in UHBD. This uniform random number generator is a combination of a Fibonacci sequence and an arithmetic sequence.

The Box-Muller technique is currently used to generate the Gaussian random numbers. For uniform random variables u_1 and u_2 on $(0, 1]$ one needs to calculate

$$g_1 = \left(\sqrt{-2\ln u_1}\right)\cos 2\pi u_2, \ g_2 = \left(\sqrt{-2\ln u_1}\right)\sin 2\pi u_2. \tag{18}$$

The numbers g_1 and g_2 are normally distributed random numbers:

$$f_g(x) = \frac{\exp\frac{-x^2}{2}}{\sqrt{2\pi}} \tag{19}$$

This method is easily implemented, but is computationally intensive.

Marsaglia's method for the generation of uniform random numbers has been parallelized and used in UHBD. The main problem is in simplifying the Gaussian transform. The objective is to use a linear transform to get random numbers with approximately a Gaussian distribution (Equation 19). Equations 18 are not linear because they contain sin, cos, log.

The distribution function $f_g(x)$ is normalized ($\int_{-\infty}^{+\infty} f_g(x)dx = 1$). Let us divide the domain $(-\infty, +\infty)$ using set of points $\{0, \pm x_1, \pm x_2, ... \pm x_{N/2-1}\}$ as delimiters such that

$$\int_0^{x_1} f_g(x)dx = \int_{x_i}^{x_{i+1}} f_g(x)dx = \int_{x_{N/2-1}}^{+\infty} f_g(x)dx = \frac{1}{N}. \tag{20}$$

If N is sufficiently large then $f_g(x)$ is approximately constant in each subdomain. We use the following method to calculate x_i.

$$x_0 = 0, \ x_{i-1/2} = x_{i-1} + \frac{1}{Nf(x_{i-1})}, \ x_i = x_{i-1} + \frac{2}{Nf(x_{i-1/2})}, \ i = 1, ... N/2. \tag{21}$$

The routine that calculates x_i is called only once during a BD simulation.

It is now necessary to approximate the Gaussian transform. Consider a uniform random number u on $(0, 1]$. The number $i = \text{Int}(Nu) - N/2$ is an integer uniform random number and can be used to determine the subdomain. The sign of i will correspond to the sign of g. Then $y = Nu - \text{Int}(Nu)$ will be a uniform random variable on $(0, 1]$, not correlated with i. Finally the Gaussian random variable is

$$g = \text{sign}(i)(x_i + (x_{i+1} - x_i)y). \tag{22}$$

This transform can be made as accurate as desired by choosing an arbitrarily large number N, without significant influence on computational time.

With the above optimization the random number generator became more than 20 times faster without any loss in accuracy.

Parallel Performance

The experiments were carried out on the Intel Delta with 16MB of memory on each node. The initial experiment was a test case for which an analytic solution is known: a single-atom target molecule with 100^3 grid size. All CPU times were computed using the dclock() system routine on the Intel system.

Figure 6 provides the total CPU time for PICCG and BD calculations with 5000 trajectories as a function of the number of processors P. It is clear that vector multiplication time, matrix action time and preconditioning time are almost scalable. Reduction time, which measures global summation for vector multiplication, increases proportionally to $\log(P)$, but is still much smaller than other terms. An algorithm can be used whose cost does not increase with with P (24). For large P, this will yield slightly improved overall performance.

The most negative effect of the parallelization on distributed memory architecture, is data exchange time. This term is proportional to the number of neighbor processors which need to be involved to exchange, which is equal $\text{Int}(P/n) + 1$,

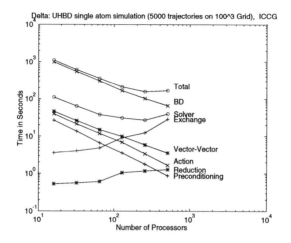

Figure 6: Total CPU time for PICCG and BD calculation with 5000 trajectories and a single-atom target molecule on a 100^3 grid as a function of the number of processors on Delta

where the mesh size s $N = n^3$. The last term dominates over others when $P > n$. For large P, a more complex decomposition should be used (the data presented here are based on strip decompositions) (24). This will have a substantial effect on the overall performance for large P.

Figure 7 demonstrates the advantage of our parallel ICCG solver for the electrostatic potential over the simpler diagonal prescaled preconditioned conjugate gradient method (DCG). Since PICCG converges twice as fast as DCG the first one is preferable.

The Figure 7 also demonstrates that increasing the number of processors initially slows down the convergence of the parallel ICCG. The reason is that the original (single processor) IC preconditioner (Figure 3) is closer to A^{-1} than the parallel IC preconditioner (Figure 5). The most surprising result which can be seen in Figure 7 is that convergence improves for both DCG and PICCG for large numbers of processors. This improvement can be explained by the fact that for large numbers of processors, each processor performs local sumations on smaller sets of floating point numbers and hence accumulates smaller round-off errors.

Figure 8 shows the difference of minimum and maximum CPU time for the electrostatic and Brownian dynamics computations. We see that the Brownian dynamic phase is more poorly load balanced than the electrostatic phase. The reason is that different trajectories have different lengths. When the total number of trajectories is small, different processors spend widely varying amounts of time doing BD (i.e. bad load balance). Parallel Brownian dynamics is most efficient when the number of trajectories is much larger than the number of processors so that all processors do approximately the same amount of work.

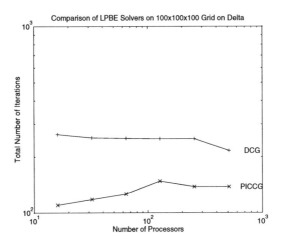

Figure 7: Total number of iterations for PICCG and DCG for a single-atom molecule on a 100^3 grid as a function of the number of processors on Delta

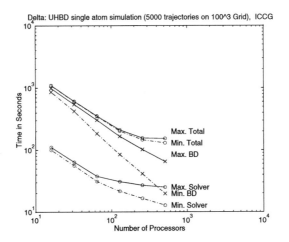

Figure 8: Total CPU time (including maximal and minimal time) for PICCG and BD calculation with 5000 trajectories and a single-atom target molecule on a 100^3 grid as a function of the number of processors on Delta

Sample Applications

The second experiment was for a case of biomedical interest and involved an enzyme thought to be involved in Alzheimer's disease, as well as the target for biological nerve agents. The enzyme studied was Acetylcholinesterase (AChE), from the electric fish (*Torpedo californica*). This enzyme is found in cholinergic synapses and catalyzes the hydrolysis of the neurotransmitter acetylcholine (ACh) into the acetate ion and choline. This process represents the termination of nerve signal transduction. The main feature of choline is the presence of a positively charged functional group (a quarternary ammonium ion). AChE has been shown to operate near the diffusion controlled limit so Brownian dynamics simulations of the substrate, ACh, approaching the enzyme should reveal any salient features of the effect of the electrostatic environment around the enzyme on diffusion of the substrate.

The X-ray structure for AChE is known and has been shown to be a homodimer which contains two active sites, one per identical monomer (25). The entire system consists of 10,416 atoms which includes hydrogens on the heteroatoms. The protonation states of the ionizable residues in AChE were determined with a procedure developed for UHBD in which each ionization state is computed based on the electrostatic environment for that particular amino acid residue in its position in the enzyme.

The electrostatic potential due to the enzyme in a dielectric continuum of water ($\varepsilon = 80$) is presented in Figure 9. A grid size of 65^3 and a grid spacing of 2.8 Å were required in order for the grid map to be compatible with the version of the display program that we used (GRASP (26)). The electrostatics part of the calculation used about 9 cpu seconds and needed 90 iterations while using 16 nodes. The entrances to the two active sites are clearly identifiable by the solid black electrostatic potential surfaces for each monomer at -2.5 kcal/mol·e. Note that the active sites are oriented in opposite directions so as not to directly compete for the substrate. Electrostatic potential contour lines are shown at -2.0, -1.25, and -0.75 kcal/mol·e to highlight the fact that the potential extends far from the active sites and effectively steers the positively charged ACh substrates to the active sites.

A sample Brownian dynamics trajectory is shown in Figure 10. In a typical Brownian dynamics experiment, many such trajectories are run and checked to see whether they satisfy the specified conditions for reaction. In the present simulation, the diffusing particle (ACh) is treated as a single sphere with a unit positive charge and a hydrodynamic radius of 5 Å. The time step in the region near the enzyme is 0.02 ps; this trajectory represents 4000 ps. Note that the diffusing substrate spends most of its time under the influence of the negative region of the electrostatic potential of the enzyme active site.

Conclusions

The determination of electrostatic free energy can be accomplished efficiently on distributed memory computers. Supercomputer-level performance is obtained on only modest numbers of processors, and techniques have been identified to allow the use of hundreds of processors efficiently. The most severe constraint in using

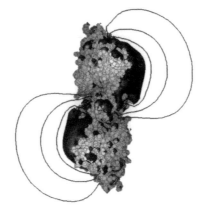

Figure 9: Electrostatic potential surface (solid black) around AChE and contoured at -2.5 kcal/mol·e. Electrostatic potential contour slices are shown for energy levels -2.0, -1.25 and -0.75 kcal/mol·e. Note that this is a homodimer with two counteropposed active sites. These data were generated with the UHBD program. The figure was generated with the GRASP software (26).

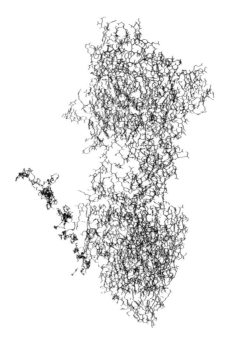

Figure 10: A sample trajectory from a Brownian dynamics run of a single sphere with a unit positive charge diffusing up to one of the active sites in AChE. The trajectory starts to the lower left of the enzyme dimer and diffuses up and into the negative electrostatic field of one of the active sites. The data were generated with the UHBD program and the figure was generated with the QUANTA software (Molecular Simulations, Inc. , Burlington, MA 01803-5297)

distributed memory computers for such calculations is the storage required for the electrostatic potential during the Brownian dynamics phase. Efficient support for distributed shared memory would remove this constraint.

Acknowledgments

This work was supported in part by grants from NIH, the Robert A. Welch Foundation, the NSF and ARPA, CRPC, and the NSF Supercomputer Centers Metacenter Program.

Literature Cited

(1) Madura, J. D.; Davis, M. E.; Gilson, M. K.; Wade, R. C.; Luty, B. A.; McCammon, J. A. *Rev. Comp. Chem.* **1994**, *5*, 229–267.

(2) Getzoff, E. D.; Cabelli, D. E.; Fisher, C. L.; Parge, H. E.; Viezzoli, M. S.; Banci, L.; Hallewell, R. A. *Nature* **1992**, *358*, 347–350.

(3) McCammon, J. A. *Current Biology* **1992**, *2*, 585–586.

(4) Luty, B. A.; Davis, M. E.; McCammon, J. A. *J. Comp. Chem.* **1992**, *13*, 1114–1118.

(5) Holst, M.; Kozack, R. E.; Saied, F.; Subramaniam, S. Multigrid Solution of the Poisson-Boltzmann Equation. Submitted for publication, 1994.

(6) Davis, M. E.; McCammon, J. A. *J. Comp. Chem.* **1989**, *10*, 386–391.

(7) Holst, M.; Saied, F. *J. Comp. Chem.* **1993**, *14*, 105–113.

(8) Il'in, V. P. *Iterative Incomplete Factorization Methods.* World Scientific: Singapore, 1992.

(9) Bagheri, B.; Ilin, A.; Scott, L. R. Parallelizing UHBD. Research Report UH/MD 167, Dept. Math., Univ. Houston, 1993. available apon request by e-mail to scott@uh.edu.

(10) Meijerink, J. A.; van der Vorst, H. A. *Mathematics of Computation* **1977**, *31*, 148–162.

(11) Meijerink, J. A.; van der Vorst, H. A. *J. Comp. Phys.* **1981**, *44*, 134–155.

(12) Berryman, H.; Saltz, J.; Gropp, W.; Mirchandaney, R. *J. of Parallel and Distributed Computing* **1990**, *8*, 186–190.

(13) Foresti, S.; Hassanzadeh, S.; Murakami, H.; Sonnad, V. *Parallel Computing* **1993**, *19*, 1–8.

(14) Hammond, S. W.; Schreiber, R. *International Journal of High Speed Computing* **1992**, *4*, 1–21.

(15) Ortega, J. M. *Itroduction to Parallel and Vector Solution of Linear Systems.* Plenum Press: New York, 1988.

(16) Rothberg, E.; Gupta, A. *Parallel Computing* **1992**, *18*, 719–741.

(17) van der Vorst, H. A. *Comp. Phys. Commun.* **1989**, *53*, 223 – 235.

(18) Bagheri, B.; Clark, T. W.; Scott, L. R. *Fortran Forum* **1992**, *11*, 20–31.

(19) Bagheri, B.; Ilin, A.; Scott, L. R. Parallel 3-D MOSFET Simulation. In *Proceedings of the* 27th *Annual Hawaii International Conference on System Sciences*, volume 1, pp. 46–54, Maui, HI, 1994.

(20) Bagheri, B.; Ilin, A.; Scott, L. R. A Comparison of Distributed and Shared Memory Scalable Architectures. 1. KSR Shared Memory. In *Proceedings of the Scalable High Performance Computing Conference*, pp. 9–16, Knoxville, TN, 1994.

(21) Almasi, G. S.; Gottlieb, A. *Highly Parallel Computing.* The Benjamin/Cummings Publishing Company Inc.: Redwood City, CA, 1994.

(22) Marsaglia, G.; Zaman, A.; Tsang, W. W. *Statistics and Probability Letters* **1990**, *8*, 35–39.

(23) James, F. *Comp. Phys. Commun.* **1990**, *60*, 329–344.

(24) Fox, G. *Solving Problems on Concurrent Processors*, volume 1. Prentice Hall: Englewood Cliffs, 1988.

(25) Sussman, J. L.; Harel, M.; Frolow, F.; Oefner, C.; Goldman, A.; Toker, L.; Silman, I. *Science* **1991**, *253*, 872.

(26) Nicholls, A.; Honig, B. *GRASP.* Columbia University: New York, v. 1.10.

RECEIVED November 15, 1994

Chapter 13

Classical and Quantum Molecular Dynamics Simulation on Distributed-Memory Massively Parallel Computers

Zhiming Li[1], R. Benny Gerber[1,2], and Craig C. Martens[1]

[1]Department of Chemistry and Irvine Research Unit in Advanced Computing, University of California, Irvine, CA 92717–2025
[2]Department of Physical Chemistry and the Fritz Haber Center for Molecular Dynamics, Hebrew University of Jerusalem, Jerusalem 91904, Israel

The implementations of classical and quantum molecular dynamics simulations on distributed-memory massively parallel computers are presented. First, we discuss the implementation of large-scale classical molecular dynamics (MD) simulations on SIMD architecture parallel computers, and in particular, on the family of MasPar distributed-memory data parallel computers. We describe methods of mapping the problem onto the Processing Elements (PE's) of the SIMD architecture, and assess the performance of each strategy. The detailed implementations of this data parallel construct are illustrated for two case studies: classical MD simulation of a two-dimensional lattice and the photodissociation mechanisms of a diatomic iodine impurity in a three-dimensional argon lattice. We also present a study of quantum dynamics using the Time-Dependent Self-Consistent Field (TDSCF) method. These calculations demonstrate the potential of using massively parallel computers in MD simulations of thermodynamic properties and chemical reaction dynamics in condensed phases.

Molecular dynamics (MD) simulation using digital computers has proved to be a useful tool for studying the properties of liquids, solid, polymers, and other condensed phase systems (1-4). In the past decade, the growing availability of fast vector and parallel supercomputers has made it possible to apply MD to increasingly more realistic and challenging problems in the fields of chemistry, biology, physics and material science (5-8). With the advent of high performance parallel computers (9-11), we face the challenge of developing new methods that make optimal use of these computational resources. In this report, we will examine the practicalities of parallelizing the basic MD algorithms on distributed-memory single instruction-multiple data (SIMD) machines, using the high performance data parallel programming language Fortran 90. The particular hardware used belongs to the MasPar family of massively parallel computers.

The potential for parallelism in many scientific applications, including molecular dynamics simulation, is primarily due to data parallelism. Here, the same operation is applied simultaneously to all elements of a large data structure.

0097–6156/95/0592–0186$12.00/0

Exploiting data parallelism on distributed memory SIMD machines requires careful partitioning of the data structure and computational tasks in order to benefit from the architecture. In this paper, we describe the parallelization of the MD force evaluation routine for many particles interacting by pairwise forces. For a system containing on the order of 1000 particles undergoing mutual pairwise interactions, our results indicate that a SIMD computer system can be a very efficient computational platform for MD simulations. The need for modification of the algorithms to achieve optimal performance and scalability for very large systems is also discussed.

The remainder of this paper is organized as follows. First we briefly describe the machine characteristics of the MasPar MP-2 massively parallel computer. Then we review the classical and quantum molecular dynamics methods. The detailed implementation of each algorithm on the MP-2 and the CPU performance results are described and followed by a discussion of the results and future work.

The MasPar MP-2 Massively Parallel Computer System

Hardware Overview. The MasPar Computer Corporation MP-2 is a fine-grain massively parallel computer. It uses a single instruction-multiple data (SIMD) architecture. Here, each processor element executes the same instruction, broadcasted by a processor array control unit (ACU), simultaneously, on its unique set of data. The MP-2 has from 1024 to 16,384 32-bit processing elements (PE's). Each PE contains a 32-bit ALU and floating point (FP) hardware to support IEEE 32- and 64-bit arithmetic, and has a local memory of either 16 or 64 Kbytes. Aggregate double precision peak floating point performance for a 4K processor machine is 600 Mflops.

The interconnection scheme for the processing elements in the MP-2 consists of a two-dimensional mesh with toroidal wraparound. There are two distinct types of communication patterns supported by the hardware. Local grid-based patterns are supported through direct use of a two-dimensional torus grid with 8-way nearest-neighbor communication (X-Net), while general patterns are supported through the global router, which implements arbitrary point-to-point communication. The general purpose network provides a bandwidth in excess of 1 Gigabyte/sec, while the X-Net provides an aggregate bandwidth exceeding 20 Gigabytes/sec in a 16K processor system.

The MasPar MP-2 is controlled by a serial front-end DECstation 5000 model 240. Application programs are compiled, debugged, and executed on the front-end computer, passing MP-2 instructions to the ACU as appropriate. To facilitate the storage of large amounts of data for data-intensive applications, the MP-2 is also equipped with a disk array system, which is NFS mountable and has 11 Gbytes capacity of formatted data with a sustained 12 MB/sec data transfer rate.

Programming Model and Environment Overview. The programming model for the MasPar massively parallel computer is data parallel computing, which means that the same operations are performed on many data elements simultaneously. Each data element is associated with a single processor. Applications are not restricted, however, to data sets matching exactly the physical size of the machine. In general, data parallel applications often require many more individual processors than are physically available on a given machine. The MasPar system provides for this through its virtual-processor mechanism, supported at the MPFortran (MPF) (12) level, and is transparent to the programmer. MPF is based on the Fortran 90 ISO standard (13). If the number of processors required by an application exceeds the number of available physical processors, the local memory of each processor is split into as many layers as necessary, with the processor automatically time-sliced among layers. If V is the number of virtual processors and P is the number of physical processors, each physical processor would support V/P virtual processors. The ratio V/P is called the virtual-processor, or VP, ratio (14,15). The concept of parallel virtuality is achieved by using both optimizing compiler technology and architecture design, in contrast with the Thinking Machine CM-2 implementation (16), which depends only on architecture design. MPF, the MasPar implementation of Fortran 90, allows the user to easily handle very large data sets. With MPF, the MP-2 massively parallel architecture is completely transparent. While working with a single application program, the MPF optimizing compiler automatically separates scalar and parallel code, assesses the MP-2's different functional units as needed, and integrates all communications and I/O operations.

Another very useful tool is the MasPar Programming Environment (MPPE) (17), which provides an integrated graphical interface environment for developing application programs.

MasPar Math and Data Display Library. The MPML (MasPar Mathematics Library) (18) consists of a set of routines for the implementation of data parallel mathematical operations. It contains three primary groups of routines: solvers, linear algebra build blocks, and fast Fourier transforms (FFTs). The solvers include a dense matrix solver, a Cholesky solver, out-of-core solvers, and conjugate gradient solvers. The linear algebra building block routines include versions that operate on blocks of memory layers, x-oriented vectors, y-oriented vectors, and matrices. The FFT routines include both real and complex versions in most cases. The eigensolvers are under development, and are expected to be released with the new version of system software.

Theoretical Background of the Molecular Dynamics Method.

Classical Molecular Dynamics Simulation. Classical molecular dynamics simulations are based on a numerical solution of Hamilton's equations of motion (19):

$$\frac{d\vec{r}_i}{dt} = \nabla_{\vec{p}_i} H = \frac{\vec{p}_i}{m_i}$$

$$\frac{d\vec{p}_i}{dt} = -\nabla_{\vec{r}_i} H = \vec{F}_i \, , \quad i=1, \ldots, N, \tag{1}$$

where H is the system Hamiltonian, which describe the time evolution of the Cartesian coordinates $\vec{r}_i$ and momenta $\vec{p}_i$ for a collection of N particles with masses m_i. The computationally most intensive step in an MD simulation is the evaluation of the forces:

$$F_i = -\nabla_i V \, , \quad i=1, \ldots, N \tag{2}$$

which are the gradients of the potential function $V(\vec{r}_1, \cdots, \vec{r}_N)$ describing all the interactions present in the system. The interaction potential may be further divided into bonded and nonbonded interactions, and is usually modeled by analytical functions (3), although *ab initio* molecular dynamics methods have been introduced (20). Although the interaction potential functions are evaluated analytically, their computational effort nevertheless consumes more than 90% of the computer time (3). The fact that most of the computation in an MD simulation is concentrated in the evaluation of the force makes it attractive to consider implementing the MD code on a parallel computer architecture. For the case of simple pairwise additive potential evaluations on conventional serial or vector machines, a Fortran code for evaluating the forces requires a loop over the $N(N-1)/2$ distinct pairs of the N atoms. The cost of evaluating the forces and potential energy of an N-atom system when all interactions are considered thus scales as N^2, which causes a severe obstacle to the simulation of very large systems.

Trajectory Propagation by the Velocity Verlet Algorithm. To integrate the equations of motion we adopted the velocity Verlet algorithm (3). This method is based on a direct solution of the second-order Newton's equations corresponding to the pair of first-order Hamilton's equations of Eq. (1):

$$\frac{d^2 \vec{r}_i}{dt^2} = \vec{F}_i \qquad i = 1, \dots, N \qquad (3)$$

The equations for advancing the position can be written as follows (3):

$$\vec{r}_i(t + \Delta t) = \vec{r}_i(t) + \Delta t \vec{p}_i(t) / m_i + (1/2)\Delta t^2 \vec{F}_i(t) / m_i \qquad i = 1, \cdots, N$$

$$\vec{r}_i(t - \Delta t) = \vec{r}_i(t) - \Delta t \vec{p}_i(t) / m_i + (1/2)\Delta t^2 \vec{F}_i(t) / m_i \qquad i = 1, \cdots, N$$

$$(4)$$

where the momenta $\vec{p}_i(t)$ are evaluated from the following formula:

$$\vec{p}_i(t) = m_i \frac{\vec{r}_i(t + \Delta t) - \vec{r}_i(t - \Delta t)}{2\Delta t}, \qquad i = 1, \cdots, N. \qquad (5)$$

A detailed analysis of the stability of this algorithm can be found in Ref. (3). The advantage of using the Verlet algorithm instead of more sophisticated integration algorithms is its simplicity. In addition, the Verlet algorithm calls the force evaluating routine only once for each time step advanced. Gear and other predictor-corrector based integration algorithms need to call the force routine twice for each integration cycle.

Quantum Molecular Dynamics Simulation. An exact quantum dynamics study is based on solving the time-dependent Schrodinger equation (21):

$$i\hbar \frac{\partial \Psi(q_1, \cdots, q_n; t)}{\partial t} = \hat{H} \Psi(q_1, \cdots, q_n; t), \qquad (6)$$

where $\hat{H}$ is the Hamiltonian operator:

$$\hat{H} = -\sum_{i=1}^{n} \frac{\hbar^2}{2m_i} \frac{\partial^2}{\partial q_i^2} + V(q_1, \dots, q_n). \qquad (7)$$

The $(q_1, \cdots, q_N)$ are a set of vibrational coordinates and $(m_1, \cdots, m_N)$ are the masses of the particles comprising the system. Solution of this exact problem numerically is currently intractable for more than 4 particles, and approximations must be used to study the quantum dynamics of many-body systems. The approximate approach we employ is the quantum Time-Dependent Self-Consistent Field (TDSCF) method (22,23).

We now illustrate the TDSCF approach by considering a collinear model system of anharmonic oscillators. The method is based on treating each vibrational mode of the system by a single mode wave function affected by the *average interaction* with other modes (22,23). The validity of the TDSCF approximation has been discussed previously (24). Using the TDSCF ansatz, the total wave function of the system is written as a product of single mode wave functions (22):

$$\Psi(q_i, \cdots, q_n; t) \cong \prod_{i=1}^{n} \phi(q_i; t) \tag{8}$$

where $\phi_i(q_i; t)$ is the one degree of freedom wave function which describes the ith vibrational mode. The TDSCF equations of motion can be written in the following form (22):

$$i\hbar \frac{\partial \phi_k(q_k; t)}{\partial t} = \hat{h}_k^{SCF}(q_k; t)\phi_k(q_k; t) \qquad (k=1,\ldots,N) \tag{9}$$

where

$$\hat{h}_k^{SCF}(q_k; t) = -\frac{1}{2m_k}\frac{\partial^2}{\partial q_k^2} + \overline{V}_k(q_k; t) \tag{10}$$

and

$$\overline{V}_k(q_k; t) = \left\langle \prod_{j \neq k}^{n} \phi_j(q_j; t) \middle| V(q_1, \cdots, q_n) \middle| \prod_{j \neq k}^{n} \phi_j(q_j; t) \right\rangle \tag{11}$$

are the single mode TDSCF potential energy functions. Note that $\overline{V}_k(q_k; t)$ depends on q_k explicitly, and also on the dynamics of the other modes implicitly, through the states $\phi_i(q_i; t)$ with $i \neq k$ in Eq. (11). Solution of the above equations is carried out simultaneously and self-consistently for all the vibrational modes. Thus a multi-dimensional wave function is reduced to a product of one-dimensional wave functions.

Wave Packet Propagation by the Grid Method. The wave packet propagation procedure used is adapted from several existing grid methods (25-28). The single mode wave function is discretized, meaning that they are represented by their values at a set of one dimensional grid points. The grid used here is a one-dimensional lattice in the coordinates q and the domain are chosen to span the dynamically relevant range of coordinate space. To calculate the time evolution of the wave function, it is necessary to evaluate the SCF Hamiltonian operator $\hat{h}_i^{SCF}\phi_i(q_i; t)$, which can be decomposed into potential energy and kinetic energy operators. The action of potential operator V on ψ is local, and its effect on the wave function is simply multiplication at each of the discrete grid points. The operation of the kinetic operator $\hat{T}\phi_i(q_i; t)$ is evaluated by the Fourier method (23,25-28). The method involves two fast Fourier transforms. First $\phi_i(q_i; t)$ is

Fourier transformed to $\overline{\phi}_i(p_i;t)$ in momentum space, then multiplied by $-\hbar^2 p_i^2 / 2m_i$ and the products are inverse Fourier transformed back to the coordinate space.

Time propagation is accomplished by the second order differencing (SOD) scheme (23). The SOD scheme can be expressed as:

$$\psi(t + \Delta t) = \psi(t - \Delta t) - 2i\widehat{H}\psi(t), \tag{12}$$

where the operation of the Hamiltonian on a wave packet is evaluated by the FFT method as discussed above. For a reasonably small time step, this procedure preserves both the norm of ψ and its energy. The detailed stability analysis of this wave packet propagation scheme was given in Ref. (25).

In summary, the TDSCF algorithm involves propagation of the single mode wave functions for all the TDSCF modes using the SOD scheme under the mean field potential function as defined in the equation (11). After each time step, the mean field potential functions are updated using the new set of single mode wave functions. This procedure is continued until a desired number of time steps is reached. The time-consuming part of a TDSCF code includes two routines: (i) the evaluation of the effect of the Hamiltonian on the wave function and (ii) the evaluation of mean field potential functions. If the grid size is M and the number of modes is n, for a serial computer these two steps scale as $n \, ln \, M$ and Mn, respectively.

Data Parallel Implementations of Molecular Dynamics Simulation

The successful implementation of classical MD and quantum TDSCF methods on SIMD computers involves a number of steps. First, variables must be allocated to either the front end or to the DPU (data parallel units). As a rule, large arrays are usually stored on the DPU and all others on the front end. To save communication costs, certain 1-D arrays are converted into 2-D arrays using the SPREAD construct. Second, the program must be designed to minimize data flow between the front end and the DPU. Finally, Fortran 90 array constructs, such as SUM and PRODUCT and machine-dependent routines, such as the MasPar fast Fourier transform routines, should be employed where possible to increase the efficiency of operations on the data.

Classical Molecular Dynamics Simulation. To overcome the unfavorable scaling of computational cost with particle number associated with serial computers, several parallel implementations of MD have been developed. We now describe two alternative decomposition strategies which we have used in our studies: mapping of each atom onto each virtual PE, and mapping each interaction (i.e., unique *pair* of atoms) onto each virtual PE. In this paper, we have implemented these strategies using MPF's data parallel constructs, which are data parallel extensions of standard Fortran (12).

Mapping each atom onto each virtual PE. There are several applications that suit the MasPar MP-2 architecture very well. These systems include two-dimensional (2D) lattice systems, which can be used to represent surface phenomena, 2D Ising models (29) and 2D Cellular Automata (30-33). For general 2D lattices, the Fortran 90 array construct CSHIFT, (which uses X-Net communication) provides a very fast means of evaluating the forces and potential functions. Using the CSHIFT construct, the distance between nearest atoms along the x Cartesian direction can be written as

$$dx = \text{CSHIFT}(rx,\text{dim}=1,\text{shift}=1) - rx, \tag{13}$$

where rx is the array of x positions of each atom. The same equation also applies to the calculation of the separations of nearest-neighbor atoms in y Cartesian direction. For long-range interaction beyond the nearest-neighbor atoms, equation (13) can be generalized as:

$$dx = \text{CSHIFT}(rx,\text{dim}=1,\text{shift}=m) - rx, \tag{14}$$

where m is number of lattice spacing needed to be included in the interaction potential functions. The force between atoms depends on the corresponding dx and dy. The total potential energy can then be obtained by the MPF construct SUM, for which computational cost scales as $\log N$, as compared to N in the sequential construct, where N is the number of elements to be summed.

Mapping each pair of interactions onto each virtual PE. The mapping scheme presented above is suitable for simulations of lattice systems—that is, systems with static and permanent particle positions. For an arbitrary N-body system with simple 2-body interactions, we use a different mapping scheme. Here, we map each *pair* of interactions onto each virtual processor element. We have implemented this mapping scheme in our MD simulation code using an MPF data parallel construct. This approach has been used previously on other SIMD systems. It was first carried out on the DAP (ICL Distributed Array Processor) (34) and later implemented on the Thinking Machine CM-2 (35). We will follow their notation while presenting the details of our implementation.

The MPF data parallel construct SPREAD provides the capacity of mapping the pair distance calculation over all pairs into "scalar-like" vector and matrix computation. For example, to calculate the x-component of the separation between all pairs of atoms, we use

$$dx = \text{SPREAD}(x,\text{dim}=1,\text{copies}=N) - \text{SPREAD}(x,\text{dim}=2,\text{copies}=N) \tag{15}$$

where dx is a 2D array of size $N \times N$, N is the number of atoms, and x is a vector of length N consisting of the x-component of atomic Cartesian coordinates. The MPF's SPREAD function broadcasts the specified element (x) into the array dx along a specific axis. Thus, the operation shown in equation (15) constructs the matrix of x_i-x_j values. Similar constructs are needed for the other Cartesian directions, and the matrix of interatomic separations r is then obtained by

$$r = \text{DSQRT}(dx*dx+dy*dy+dz*dz), \tag{16}$$

which also represents the element-by-element operations implied by the MPF array extension. Potential energy and interatomic forces are evaluated as their full $N \times N$ matrix values and then they are summed to yield vector forces and scalar potential energy using the SUM function within MPF.

Quantum Molecular Dynamics Simulation. The TDSCF algorithm combined with the discretized representation of the single mode wave functions is an ideal system for the SIMD massively parallel computer architectures. Here we present the detailed implementation. The essence of the approach is to map each grid point of the wave function to each virtual PE. We discretize the single mode wave function on a one dimensional lattice of M grid points, and assume the system consists of n TDSCF modes. Thus, the total wave function of the system can be represented as a two-dimensional array of rank $M \times n$. That is:

$$\phi(i,j) \qquad q_j^i = i\Delta q, \qquad i = 1, \cdots, M, \qquad j = 1, \cdots, n, \qquad (17)$$

where Δq is the grid spacing, M and n are the number of grid point and the number of TDSCF modes, respectively. A similar decomposition scheme has been used in quantum scattering calculations (36).

One of the most time-consuming parts of the TDSCF approach is the evaluation of the mean-field potential given in Eq. (11). This involves numerically evaluating the potential integral over the n-dimensional configuration space of the system. For our model system, such integrals reduce to a function of one dimensional integrals, which are approximated by the trapezoidal rule (37).

Timings and Performance Measurements

Newton's equations of motion were integrated numerically using the velocity Verlet algorithm (3). The following CPU benchmark measurements are for 400 cycles of integration.

Anharmonic 2D Lattice Model System. The MP-2204 is a 4096 processor massively parallel computer. The processors are arranged on a 64x64 mesh, which provides fast nearest-neighbor communication. The machine works well for explicit algorithms that take advantage of this architecture. As we mentioned above, a 2D lattice system fits this architecture very well. In our model calculations, we use a 2D Morse lattice. The Morse potential has the following form:

$$V(R) = D_e [1 - \exp(-\beta(R - R_e))]^2, \qquad (18)$$

where R is interatomic distance, and D_e, β and R_e are potential parameters. For distances greater than a cutoff $R > R_C$, we set the potential to zero; the potential outside of this radius does not need to be evaluated. In our performance studies, we vary R_C to determine scaling of computational effort with cutoff radius. The detailed implementation involves using the MPF construct CSHIFT, which uses fast X-Net communications.

The computational time needed per time step of the molecular dynamics simulation on this system is approximately given by:

$$CPU = AN + BNM^2 \qquad (19)$$

Here, A is the computing time needed per molecule in the Verlet propagating step and B is the time needed per X-Net communication for each pair of particles. M is the number of lattice spacings included in the force evaluation, which depends in turn on R_C. For system with the nearest-neighbor interactions $M = 1$, second nearest-neighbor $M = 2$, etc.

The scaling of computer time with problem size and number of PE's are important performance indicators for massively parallel computers. Here, the scalability of the MD code is tested with respect to both the number of processor elements (PE's) and the number of atoms. For the 2D lattice system, the CPU performance of the parallel program versus the number of atoms is shown in Figure 1. The calculation was done on the MP-2204. A linear dependence on the number of atoms is observed. The peak CPU performance for MP-2204 is 600 Mflops for double precision floating point algorithms. The parallel version of this program running on the MP-2204 can reach a speed of 300 Mflops, which is three times faster than a single processor of the Cray Y-MP. In this particular application, we have achieved 50% of the MP-2's peak performance.

M depends linearly on R_C, and from Eq. (15), we can see that the computational time is proportional to the square of the interaction cut-off radii of the potential function. This expected behavior is in fact observed, as indicated in Figure 2.

Photodissociation Dynamics of I₂ in Rare Gas Solids. The system considered in this study consists of a single I₂ molecule embedded in a double substitutional site of a face-centered cubic Ar lattice, consisting of a cube of 512 atoms with periodic boundary condition. The interaction potential function for the system is given by:

$$V = V_{I_2} + V_{Ar} + V_{I_2-Ar},$$ (20)

where the individual terms correspond to the I-I interaction potential, the Ar-Ar interaction potential, and the interaction potential between the I₂ and the Ar atoms, respectively. For ground state I₂, we use a Morse potential to model the I-I interaction:

$$V_{I_2}(R) = D_e[1 - \exp(-\beta(R - R_e))]^2,$$ (21)

where D_e is the dissociation energy of ground (X) electronic state, β is the potential range parameter, and R_e is the equilibrium distance of I₂. Photodissociation is modeled in this system by an instantaneous transition from the X ground state to the C repulsive excited electronic state. The I₂ excited state potential function is given by:

$$V_{I_2}(R) = A \exp(-\alpha R),$$ (22)

where R is the instantaneous I-I distance and the parameters A and α are obtained by fitting spectroscopic data (38). The nonbonded Ar-Ar and Ar-I interactions are modeled by pairwise Lennard-Jones potentials, given by:

$$V(r) = 4\varepsilon\left[\left(\frac{\sigma}{r}\right)^{12} - \left(\frac{\sigma}{r}\right)^6\right].$$ (23)

The potential parameters are given in Table 1. The photodissociation dynamics of I₂ in solid Ar were treated using classical mechanics (39,40). To prepare the initial conditions for the photodissociation, the system was initially equilibrated at 15 K with I₂ in its ground electronic state. The positions and velocities of each atoms were written out to a file every 50 fs. The configurations and velocities were read in from this file as the initial conditions for ensemble of trajectories. The photodissociation process was then simulated by suddenly switching the I₂ potential from its ground state Morse form to the exponential form of the excited repulsive potential surface.

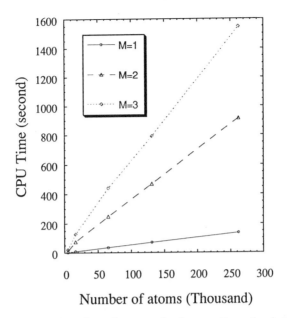

Number of atoms (Thousand)

Figure 1. CPU benchmark performance for the two dimension lattice with nearest-neighbor interaction on MP-2(2204). The plot shows the CPU time versus number of atoms. Note that all the timing results are for the CPU time of integrating 400 cycles using the velocity Verlet algorithm.

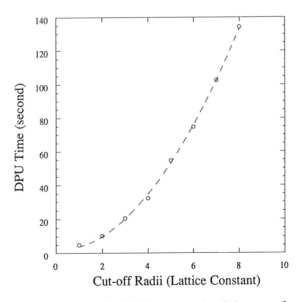

Cut-off Radii (Lattice Constant)

Figure 2. Data Parallel Unit (DPU) computational time as a function of interaction cut-off radii for the simulation of 2d lattices with 4096 (64x64) particles.

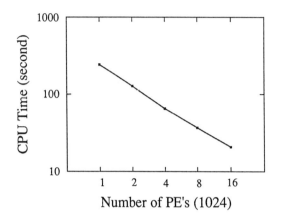

Figure 3. CPU time of benchmark calculation for the parallel program versus the number of processor elements used on MP-1 for system with 512 atoms.

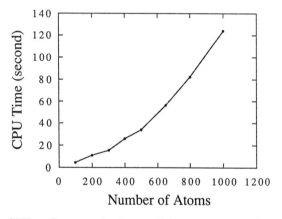

Figure 4. CPU performance for the parallel program versus the number of atoms on MP-2204 for an N-body system.

Table 1. Potential parameters.

Potential Parameters	Values
D_e	15,370 cm^{-1}
r_e	2.55 Å
β	1.77 Å^{-1}
A	5.10 x10^7 cm^{-1}
a	3.29 Å^{-1}
ε_{Ar-Ar}	83.26 cm^{-1}
σ_{Ar-Ar}	3.405 Å
ε_{I-Ar}	130.24 cm^{-1}
σ_{I-Ar}	3.617 Å

To obtain optimal performance on a SIMD massively parallel computer, we need to keep the operations performed on the data as regular as possible. This is not a problem for a homogeneous system like pure Ar. For an atomic system with a molecular impurity, such as I_2 in solid Ar, there are three different kinds of interactions, as indicated in Eq. (20). To make effective use of the SIMD architecture, we use masking operations for each type of interaction. For interactions between Ar atoms, the Lennard-Jones potential parameter ε is represented by an $N \times N$ matrix with following matrix elements: $\varepsilon_{ij} = \varepsilon_{Ar-Ar}$ if i and j represent the Ar atoms and $\varepsilon_{ij} = 0$ otherwise. Similar matrices are constructed for the interactions between Ar and I, and for the Morse oscillator parameters representing the interaction between I atoms. The potential energy and force evaluations can be written in a very concise form, and the force evaluation can be implemented more efficiently, but at the expense of computing terms (albeit in parallel) which do not contribute to the interparticle forces.

The dependence of the computational time on the number of particles can be written as:

$$CPU = AN + CN \quad \text{if } N(N\text{-}1)/2 < P,$$

$$CPU = AN + CN^2 \quad \text{if } N(N\text{-}1)/2 > P,$$

(24)

where N is the number of particle in the system and P is the number of processors. For an N-body system with all pair interactions, the benchmark calculation of the parallel code versus the number of atoms simulated on MP-2204 is shown in Figure 3. In this calculation, we have mapped each pair of interactions to a PE. The number of pairwise interactions is $N(N\text{-}1)/2$, where N is the number of atoms. For MP-2204, there are 4K PE's. For systems with more than 64 atoms, the parallel virtual concept, which uses the PE memory as virtual processors, is employed. The maximum number of atoms one can include in the simulation is limited by the size of local PE memory. For the MP-2204 with 64 KBytes of local memory, we can simulate up to 1200 atoms. When using virtual PE's, the relation between the CPU time and the number of atoms is no longer linear.

We now consider the scalability of the CPU performance with respect to the number of PE's. Figure 4 shows the CPU time of the MP-1 versus the number of PE's for a 512 atom system with all pairwise interactions included. A nearly linear relation was obtained, which indicates very good scalability performance.

Table 2. The CPU benchmark results of the classical MD code of 512 atoms with all pair interactions for 400 cycles.

Computer	CPU Time (second)	Mflops	Peak Mflops	% Peak
MP-1204	63	62	128	48
MP-1208	32	124	275	45
MP-2204	23	172	600	29
MP-2216	7.7	510	2400	21
Cray Y-MP (unicos)	13	151	330	46
IBM ES9000 (AIX)	74	26.5	50	53
DEC Alpha 3000/400	102	19.2	100	19
SGI Indigo (R4000)	185	10.6		
Convex C240	207	9.5	50	19

Table 3. The CPU benchmark results of the quantum molecular dynamics simulation (TDSCF code) of 17 atoms with a grid size of 256 integrated for 100 cycles.

Computers	CPU (second)	Mflops	Peak Mflops	% Peak
MP-2204	23	172	600	29
MP-2216	7.7	510	2400	21
Intel Touchstone Delta*	16	255	640	39
Cray Y-MP (unicos)	13	151	330	46

* The timing result is for a 16-node partition.

Quantum System: Molecular Hydrogen Clusters. We now apply the TDSCF approach to a model system consisting of a collinear $(H_2)_{17}$ cluster to demonstrate the potential use of massively parallel computers in many-body quantum simulations. The system under consideration consists of a collinear chain of 17 particles with nearest-neighbor interactions. We simulate the dynamics of the para-hydrogen species; i.e., H_2 is in the rotational ground state (J=0). The ratio of the minor to the major axis of the ellipsoidal electronic charge distribution of the H_2 molecule is very close to unity (41). Thus, molecular para-hydrogen can be represented, for simplicity, as structureless spheres. We denote by x_i the Cartesian coordinate of the atom i and by r_{eq} the equilibrium distance between neighboring atoms. The interaction potential between H_2 molecules is modeled by a Morse potential function. Thus, the total potential function of the model cluster system is

$$V(x_1,\cdots,x_n) = \sum_{i}^{n-1} D_e[1 - e^{-\beta(x_{i+1}-x_i-r_{eq})}]^2. \tag{25}$$

The Morse potential parameters are D_e=32.39K, β=1.65Å^{-1}, and r_{eq}=3.44Å. These parameters were obtained by fitting the more accurate but complicated potential function available from the molecular beam data (42) and *ab initio* electronic structure calculations (43-45).

The dependence of the computational time on the number of grid points and number of modes can be expressed as follows:

$$CPU = A \ln M \qquad \text{if } Mn < P,$$

$$CPU = \frac{An}{P} \ln M \qquad \text{if } Mn > P, \tag{26}$$

where M is the number of grid points for the wave function discretization, n is number of TDSCF modes in the system and P is the number of processors. These expressions result from the fact that both fast Fourier transforms and summations over PE's, which are required in the evaluations of the Hamiltonian operation and the TDSCF mean field potential functions, scale as $\ln M$.

Comparison with the Cray Y-MP and Other Computers. Table 2 shows the benchmark calculation of the 512 atom system on a selection of computers. The implementation on the vector machines was coded in a highly vectorized form in Fortran 77. The data parallel version of the code on MP-2216 ran 1.7 times faster than the Y-MP for this problem, and was estimated to be running at about 250 Mflops.

MIMD Computer implementation. The TDSCF approach was also implemented on a MIMD computer, the Intel touchstone Delta. Here we assign each single mode wave function to a node. The communication required for evaluating the mean field potential function was done using an explicit message passing scheme. The timing results for both the SIMD and MIMD implementations of the TDSCF approach are listed in Table 3.

Summary and Future Work

In this paper, Parallel versions of classical and quantum molecular dynamics simulation codes on the MP-2 were described, which achieve a substantial speed-up over a sequential version of the program on conventional vector or scalar machines.

The classical MD code is used routinely for production runs in our studies of rare gas liquid and photodissociation process in solid state materials. The performance of 1.7 times of a single processor of Cray Y-MP is obtained. Aided by the parallel constructs build into the data parallel programming language Fortran 90, we find the massively parallel computer to be a powerful research tool for molecular dynamics simulation.

In our present implementation of mapping pairwise interactions onto each PE, the system size (i.e., number of atoms) amenable to simulation is limited to being less than $N = 1300$ on the MP-1208. Alternative mapping schemes need to be used for simulating very large systems. One of these mapping schemes is based on associating one atom to each PE and using the Verlet neighbor-list or link cell method to keep track of interacting atoms (46). Efficient parallelization of neighbor-list generation is a problem which warrants future study.

Acknowledgments. We would like to thank Prof. V. A. Apkarian and Prof. I. Scherson for helpful discussions. CCM acknowledges support by the National Science Foundation and the Office of Naval Research. RBG acknowledges support by the US Air Force Phillips Laboratory (AFMC) under the contract F29601-92K-0016. The Irvine Research Unit in Advanced Computing is supported by the National Science Foundation. We also thank the UCI Office for Academic Computing for technical assistance, and the MasPar Computer Corporation for partial support through the allocation of computer time on the MasPar MP-2204 and MP-1208 computers.

References

1. *Simulation of Liquids and Solids: Molecular Dynamics and Monte Carlo Methods in Statistical Mechanics*; Ciccotti, G.; Frenkel, D; McDonald, I. R., Eds.; Elsevier Science: New York, NY, 1987.
2. Hockney, R. W.; Eastwood, J. W. *Computer Simulation Using Particles*; Adam Higer: New York, NY, 1989.
3. Allen, M. P.; Tildesley, D. J. *Computer Simulation of Liquids;* Clarendon Press: Oxford, 1987.
4. Could, H.; Tobochnik, J. *An Introduction to Computer Simulation Methods: Applications to Physical Systems*; Addison-Wesley: Reading, MA, 1988.
5. Rapaport, D. C. *Comp. Phys. Comm.* **1991,** 62, 198.
6. Rapaport, D. C. *Comp. Phys. Comm.* **1991,** 62, 217.
7. Rapaport, D. C. *Comp. Phys. Comm.* **1993,** 76, 301.
8. *Molecular Dynamics Simulation of Statistical Mechanical Systems*; Ciccotti, G.; Hoover, W. G., Eds.; North-Holland: Amsterdam, 1986.
9. Smith, W. *Comp. Phys. Comm.* **1991,** 62, 229.
10. Abraham, F. F. *Advances in Physics* **1986,** 35, 1.
11. Finchan, D. *Molecular Simulation* **1987,** 1, 1.
12. *MasPar Fortran Reference Manual*; Maspar Computer Corporation: Sunnyvale, CA, 1991.
13. Metcalf, M.; Reid, J. *Fortran 90 Explained*; Oxford University Press: New York, NY, 1990.
14. Blolloch, G. E. *Vector Model for Data-Parallel Computing*; MIT Press: Cambridge, MA, 1990.
15. Lewis, T. G.; El-Erwini, H. *Introduction to Parallel Computing,* Prentice-Hall: Englewood Cliffs, NJ, 1992.
16. Hillis, W. D. *The Connection Machine*; MIT Press: Cambridge, MA, 1985.
17. *MasPar Parallel Programming Environment*; Maspar Computer Corporation: Sunnyvale, CA 1991.

18. *MasPar Mathematics Library Reference Manual;* Maspar Computer Corporation: Sunnyvale, CA 1992.
19. Goldstein, H. *Classical Mechanics,* 2nd ed. Addison-Wesley: Reading, MA, 1980.
20. See for example, Car, R.; Parrinello, M. *Phys. Rev. Lett.* **1985,** 55, 2471.
21. Landau, L. D.; Lifshitz, E. M. *Quantum Mechanics,* Pergamon Press: New York, NY, 1977.
22. Gerber, R. B.; Ratner, M. A. *J. Phys. Chem.* **1988,** 92, 3252.
23. Gerber, R. B.; Kosloff, R.; Berman, M. *Comp. Phys. Rep.* **1986,** 5, 59.
24. Alimi, R.; Gerber, R. B.; Hammerich, A. D.; Kosloff, R.; Ratner, M. A. *J. Chem. Phys.* **1990,** 93, 6484.
25. Tal-Ezer, H.; Kosloff, R. *J. Chem. Phys.* **1984,** 81, 3967.
26. Kosloff, D.; Kosloff, R. *J. Comput. Phys.* **1983,** 52, 35.
27. Kosloff, R.; Kosloff, D. *J. Chem. Phys.* **1983,** 79, 1823.
28. Bisseling, R.; Kosloff, R. *J. Comput. Phys.* **1985,** 59, 136.
29. Chandler, D. *Introduction to Modern Statistical Mechanics*; Oxford University Press: New York, NY, 1987.
30. *Cellular Automata and Modeling of Complex Physical Systems*; Manneville, P.; Bccara, N.; Vichniac, G. Y.; R. Bidaux, R., Eds.; Springer Proceedings in Physics 46; Springer-Verlag: Heidelberg, 1989.
31. Vichniac, G. *Physica D* **1984,** 10, 96.
32. Stauffer, D. *Physica A* **1989,** 157, 654.
33. Gerling, R. W. In *Computer Simulation Studies in Condensed Matter Physics III*; Landau, D. P.; Mon, K. K.; Schutler, H.-B.; Springer-Verlag: Heidelberg, 1991.
34 Boyer, L. L.; Pawley, G. S. *J. Comp. Phys.* **1988,** 78, 405.
35. Brooks, C. L.; Young, W. S.; Tobias, D. J. *The International Journal of Supercomputer Applications* **1991,** 5, 98.
36. Hoffman, D. K.; Shrafeddin, O. A.; Kouri, D. J.; Carter, M. *Theo. Chim. Acta.* **1991,** 79, 297.
37. Press, W. H.; Flannery, B. P.; Teukolsky, S. A.; Vetterling, W. T. *Numerical Recipes: The Art of Scientific Computing;* Cambridge University Press: Cambridge, 1989.
38. Tellinghuisen, J. *J. Chem. Phys.* **1985,** 82, 4012.
39. Alimi, R.; Gerber, R. B.; Apkarian, V. A. *J. Chem. Phys.* **1990,** 92, 3551.
40. Zadoyan, R.; Li, Z.; Ashijian, P.; Martens, C. C.; Apkarian, V. A. *Chem. Phys. Lett.* **1994,** 218, 504.
41. Silvera I. F. *Rev. Mod. Phys.* **1980,** 52 393.
42. Buck, U.; Huisken, F.; Kohlhase, A.; Otten, D. *J. Chem. Phys.* **1983,** 78, 4439.
43. Wind, P.; Roeggen, I. *Chem. Phys.* **1992,** 167, 247.
44. Wind, P.; Roeggen, I. *Chem. Phys.* **1992,** 167, 263.
45. Wind, P.; Roeggen, I. *Chem. Phys.* **1992,** 174, 345.
46. Morales, J. J.; Nuevo, M. J. *Comp. Phys. Comm.* **1992,** 69, 223.

RECEIVED November 15, 1994

Chapter 14

Biomolecular Structure Prediction Using the Double-Iterated Kalman Filter and Neural Networks

James A. Lupo, Ruth Pachter, Steven B. Fairchild, and W. Wade Adams

Materials Directorate, Wright Laboratory, WL/MLPJ,
Wright-Patterson Air Force Base, OH 45433

The parallelization of the PROTEAN2 molecular structure prediction code has been completed for the Thinking Machines, Inc. CM-5. Benchmark and parallel performance analysis results are summarized and compared with those obtained on a Cray C90 using multiple processors in autotasking mode. The choice of an optimal machine is shown to be dependent on the size of the model studied.

In our continuing efforts towards the design of non-linear optical chromophore containing biomolecules *(1,2,3,4)* that enable the flexibility of controlling structure, an integrated computational approach has been developed. First, a neural network is trained to predict the spatial proximity of C_α atoms that are less than a given threshold apart. The double-iterated Kalman filter (DIKF) technique (coded in PROTEAN2 *(5)*) is then employed with a constraints set that includes these pairwise atomic distances, and the distances and angles that define the structure as it is known for the individual residues in the protein's sequence. Finally, the structure is refined by employing energy minimization and molecular dynamics. Initial results for test cases demonstrated that this integrated approach is useful for molecular structure prediction at an intermediate resolution *(6)*. In this paper, we report the parallelization and other aspects of porting PROTEAN2 to the CM-5 and the Cray C90.

Massively parallel processor (MPP) systems use a relatively new computer architecture concept that may enable significant speedup increases by allowing a single user to harness many processors for a single task. Experience has shown, however, that the suitability of any given program to parallelization is highly dependent on the problem being solved and the machine architecture. Ultimately, speedup is limited by Amdahl's law expressed as:

$$S_p = \frac{p}{\alpha(p-1)+1} \tag{1}$$

0097–6156/95/0592–0202$12.00/0
© 1995 American Chemical Society

where p is the number of processors, S is the speedup factor, and α is the percentage of computing done sequentially, i.e. on one processor *(7)*. Note that for $p\sim1$, S increases linearly, while for large p, the nonlinear effects may be dominant, even if the percentage of computing done on one processor may decrease due to the large parts of sequential code.

As a result of Amdahl's law, many programs show significant non-linear scaling behavior because they contain large parts of sequential code, so that the execution speed increases less than the increase in the number of processors. Thus, while the new MPP machines may have impressive theoretical performance figures, their actual performance is problem dependent, and care must be taken to find the best machine on which to run a problem. It also implies that considerable effort is required to find efficient parallel algorithms. In this work we discuss porting issues, while performance results are summarized that compare the CM-5 with the multiple node Cray C90.

Results and Discussion.

1) Approach. In the first stage, an expert system was used to develop the training set consisting of specific protein structures obtained from the Brookhaven Protein Data Base (PDB), with the data files being preprocessed to extract the backbone atomic coordinates and calculate the appropriate torsion angles. Secondary structural motifs are determined by searching for sequential residues whose torsions fall within a user defined tolerance. A neural network learns to predict secondary structure *(8,9)*, but moreover the spatial proximity of $C\alpha$ atoms. Tertiary structure information is generated in the form of binary distance constraints between the $C\alpha$ atoms, being 1 if their distance is less than a given threshold, and 0 otherwise. The proteins used in the training set were the first 48 of the set collected by Kabsch and Sander *(10)*, while several of the last sixteen proteins in this set were for testing (total of 8315 residues). A feed forward neural network with one hidden layer was used, and backpropagation was the learning algorithm. Details of this neural network application are described elsewhere *(11)*.

The DIKF *(12,13)* algorithm is subsequently employed to elucidate the structure using a data set that includes these pairwise atomic distances, and the distances that define the chemistry and stereochemistry of the molecular structure. It is notable that the neural network constraints set was found to be adequate compared to modeling with an exhaustive set of all $C\alpha$ pair distances derived from the crystal structure of Crambin *(14)*.

In particular, the structural molecular model of a polypeptide consisting of N atoms is described by the mean cartesian coordinates $\mathbf{x}$ and the covariance matrix $\mathbf{C(x)}$. The elements of $\mathbf{C}$ for any two atoms i, j are symmetric:

$$C(x_{ij}) = \begin{matrix} \sigma_{x_i x_j} & \sigma_{x_i y_j} & \sigma_{x_i z_j} \\ \sigma_{y_i x_j} & \sigma_{y_i y_j} & \sigma_{y_i z_j} \\ \sigma_{z_i x_j} & \sigma_{z_i y_j} & \sigma_{z_i z_j} \end{matrix} \tag{2}$$

The matrices on the diagonal of the covariance matrix ($C(x_{ii})$) describe the extent of three dimensional uncertainty in the position of the i-th atom, and the covariance between two variables (the non-diagonal elements $C(x_{ij})$) estimates their correlation. Distance and dihedral angle measurements are given by:

$$z = h(x) + v \qquad (3)$$

where z is the observed value of the data, $h(x)$ expresses the mathematical relationship between the state vector variables and the distance and dihedral angle values, and v represents the variance of the data.

Given this information, a sequential linear estimator for the minimum variance estimate of the state is obtained by the extended Kalman filter for non-linear measurement functions (13):

$$x(+) = x(-) + K[z - h(x(-))]$$
$$C(+) = C(-) - KHC(-) \qquad (4)$$

where $(-)$ signifies a previous structural representation to be sequentially updated to $(+)$. The criterion for the choice of the Kalman estimator gain matrix K, given by:

$$K = C(-)H^T[HC(-)H^T + v]^{-1} \qquad (5)$$

is to minimize a weighted scalar sum of the diagonal elements of the error covariance matrix C. The term within the inverse in the expression for K represents the variance of the observed measurement ($C(v) = C(h(x) + v)$). The first-order Taylor approximation of $C(h(x))$ is HCH^T, where H is the derivative of the data model h, and H^T is the transpose of H. The derivatives of the distance model are calculated analytically, while the derivatives of the dihedral angle data model are approximated by using a finite difference calculation. Note that in addition to the optimality of K contained in its structure, it can also be seen as the *ratio* between the uncertainty in the estimate and that of the measurement.

The extended Kalman filter approach is used to obtain higher-order non-linear filters by an iterative process:

$$x(+)_k = x(-) + K_k\{z - [h(x)_{k-1}) + H(x(-) - x(+)_{k-1})]\} \qquad (6)$$

with a similar expression for $C(+)_k$ for any iteration k. This iterative procedure is carried out for each one of the distance and dihedral angle constraints. However, since the filter is not optimal in the non-linear case ($h(x)$ in equation (3)), residual inaccuracies may still result. Therefore, the mean positions obtained after all data are introduced are used for another cycle of updating. The covariance matrix is reset to its initial large value in order to allow atoms freedom to move in response to the constraints, and all measurements are re-introduced into the system for each of these doubly-iterated cycles. The successive cycles are repeated until all of the constraints are satisfied to within a pre-set threshold of standard deviations from the error *e* given by:

$$e = \frac{h(x)_{final} - z}{\sqrt{v}} \tag{7}$$

Applications of the DIKF technique indicate that a known structure can be reproduced to within a small RMS error even when a limited data set is used *(15,16)*. Several applications using NMR NOE data utilizing this novel method proved the approach successful *(17,18,19)*. Note that the input to PROTEAN2 is automated *(20)*.

a) Crambin and BPTI Study. The neural network results for Crambin *(6)* were used for the application with the DIKF. The model system consisted of 327 atoms and pseudo-atoms, 1122 distance constraints, including bond lengths, distances implied by bond angles, non-bonded distances within well-defined secondary structural motifs, and 37 dihedral angle constraints. A comparison with the experimental X-ray structure *(12)* results in total all-atom average RMS of 2.4Å. Similarly, we have tested the genetically engineered Eglin-C *(21)*, and BPTI *(22)*. The model consists of 454 atoms, with 1550 distance and 60 dihedral angle constraints. Good convergence was obtained (Figure 1). A comparison with the experimental X-ray structure *(22)* results in a total all-atom average RMS of 3.5Å.

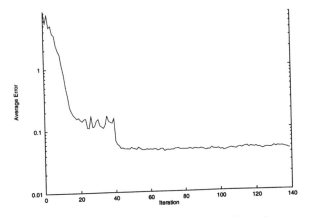

Figure 1: Average error (SD) vs. number of iterations

These results indicate that such an integrated approach may be useful for an intermediate biomolecular structure prediction, to be further refined by energy minimization and molecular dynamics. Thus, the code has been ported and evaluated the study of larger molecular systems.

2) Ports. Portability is an important software engineering issue. In the case of distributed memory systems, each architecture has its own set of system library routines which support the passing of data between processors. A program written explicitly for one architecture will have to be modified when using the library of another architecture. Given the pervasive nature of data exchange between processors, this may entail an extensive effort. The message passing code required also tends to obfuscate the programs general logic flow. These issues were resolved in this work.

A standard high-level language with embedded compiler directives is utilized. Indeed, the characteristics of the CM-5 necessitated the use of Fortran-90. The CM-5 is a distributed memory parallel processor and supports both data parallel and message passing models of parallel programming. However, the system software currently does not allow message passing programs to access the vector processing units on the processing elements. The vector units are currently supported only by the Fortran-90 compiler, which uses the data parallel programming model. The program should be portable to other systems supporting such a compiler, although the Fortran-90 version of PROTEAN2 has not yet been tested on other systems. The number of memory utilization compiler directives embedded in the source code and the dependence on the CMSSL libraries, suggests the amount of porting effort will be strongly machine dependent.

The PROTEAN2 *(5,23)* molecular structure code was ported to the CM-5 by translating it from Fortran-77 into Fortran-90. The initial port of PROTEAN2 to the CM-5 was reported earlier *(8)*. However, a system bug in the CM Scientific Support Library (CMSSL) prevented the program from running large problems on more than 32 nodes. The source code was thus gradually reduced to a set of 14 lines of Fortran which replicated the bug. Working from this demonstration program, the bug in CMSSL could be identified and was thereafter released in a new version in November 1993. A full set of benchmarking runs were then performed on the CM-5.

At the same time, experimentation with multiprocessing on a Cray C90 was underway with the fully vectorized version of PROTEAN2. This involved selection of the proper compiler options to enable autotasking, and setting up the correct environment variable (NCPUS) to specify the number of processors desired for a run. Without additional hand optimization, the autotasking support from the compiler is essentially limited to loop unrolling, with sub-loops assigned to different processors. The same series of problems run on the CM-5 were run with two C90 processors.

3) Program Performance. Two molecular models were selected for the benchmark runs, specifically the genetically engineered protein N-acetyl Eglin-C (Eglin-C), that contained 530 atoms, with a total of 1776 constraints. A second system was based on the *trp* repressor (Trp), and contained 1504 atoms and 6014 constraints *(18)*. Although both the CM-5 and the C90 systems are multi-user, the CM-5 timing functions are known to be impacted by system load. The timing figures presented should therefore be taken as indicative of typical production runs rather than the absolute best performance.

The timing figures listed by the CM-5 report the total time consumed by the service partition, while the C90 reports the total time consumed by all processors assigned to a job as well as the connect time. In the following comparisons, C90 connect time was assumed to be an equivalent metric to the total time reported by the CM-5. The C90 also reports the time that is consumed running on only one processor and the time spent running concurrently on multiple processors. This makes it easy to gauge parallel performance, but since internal timing instrumentation reports total CPU time, it also requires that internal times are converted to equivalent connect times. For a job run on two nodes, the following equations hold.

$$T_1 = \frac{(2 - C_f)T_{total}}{C_f} \tag{8}$$

$$T_2 = T_{total}\left(1 - \frac{1}{C_f}\right) \tag{9}$$

and

$$T_c = T_1 + T_2 \tag{10}$$

where T_{total} is the total C90 CPU time, T_1 is the CPU time on one node, T_2 is the concurrent CPU time on each of the two nodes, C_f is the concurrancy factor, and T_c is the C90 connect time.

(a) Timing Results. The times reported here are the times it takes to complete one full iteration cycle, the time spent in the DIKF portion of the code, and the time spent in the van der Waals correction portion (KVDW). The Eglin-C model was set up to run on 32, 128, and 256 nodes of the CM-5, while the Trp model was run on 128, 256, and 512 nodes. Both models were run on two nodes of the C90. The CM-5 times are shown in Table I while the C90 times are in Table II.

Comparing the C90 connect times with the CM-5 total times, it is observed that the two C90 nodes are 3.8 times faster than 32 CM-5 nodes on the Eglin-C problem. However, 256 nodes on the CM-5 are 2.2 times faster than two C90 nodes on the Trp problem. In fact, the single node time for the Trp model on the C90 was 6748 seconds, and the concurrent time on two nodes was 5540 seconds. The single CPU sequential time on the C90 is longer than the CM-5 time. Thus, at best, the C90 could use all 16 processors and reduce the connect time to 7095 seconds, which is 1.25 times longer than the CM-5 time. The CM-5 is clearly the faster machine for larger problems, while the C90 remains optimal for smaller problems.

Table I: CM-5 PROTEAN2 timings for one full iteration (sec)

Model	Nodes	Total Time	DIKF Time	KVDW Time
Eglin-C	32	449	177	272
	128	436	171	265
	256	492	218	274
Trp	128	6799	1031	5768
	256	7098	1417	5681
	512	8106	2216	5890

Table II: C90 PROTEAN2 timings for one full iteration (sec)

Model	Nodes	Total Time	DIKF Time	KVDW Time	Con-currancy Factor	Connect Time
Eglin-C	2	153	131	22	1.30	118
Trp	2	17790	13788	4002	1.45	12288

(b) Parallel Performance. The CM-5 Fortran compiler allows for the collection of performance profile information. Additional information can also be collected from the PRISM interactive debugger. On the C90, figures reported by the Job Accounting system allow one to compute some of the same metrics. Of particular interest is the parallel efficiency seen on the two machines. For the Eglin-C model, the C90 achieved a parallel efficiency of 46% while the CM-5 reached 66%. Both machines improved on the larger Trp model, with the C90 reaching 62% and the CM-5 reaching 82%.

Communications is an issue on the CM-5, but not on the C90 since it is a shared memory system. One performance measure considers the ratio of communications time to CPU time. For the Eglin-C model, the ratio was 0.51, while the Trp model had a ratio of 0.72. Considering the parallel efficiency figures and the communications to CPU times ratios, it is clear that the CM-5 performs better over all, but becomes communications bound. This explains the lack of scaling seen in any of the models on the CM-5. Larger problems will make better use of the CPU's at the expense of increasing communications loads.

Conclusion.

The PROTEAN2 program makes good use of the data parallel programming model on the CM-5. It does not scale well with the size of problems so far considered, but it is efficient even though communications bound. For problems of the size of the Trp model, it is clearly much preferred over the C90. The C90, however, is not without merit, and does run smaller problems much faster. Given the availability of both machines, testing a new problem on both should be done prior to conduction production runs. It should be noted that the creation of an executable autotasking version of PROTEAN2 on the C90 took less than four man hours, while nearly two man months were expended getting the Fortran-90 version to function on the CM-5. No hand optimization has been done to the C90 version, suggesting there is room for future improvements.

Acknowledgments.

This research was supported in part by a grant of High Performance computing (HPC) time from the DoD HPC Shared Resource Centers. Cray C90 time was provided by the Army Corps of Engineers Waterways Experimental Station, Vicksburg MS. This research was supported in part by the Army Research Office contract number DAAL03-89-C-0038 with the University of Minnesota Army High Performance Computing Research Center (AHPCRC). The AHPCRC Thinking Machines Inc. CM-5 timing results are based upon a beta version of the system software and, consequently, is not necessarily representative of the performance of the full version of this software.

Literature Cited.

1.   Levine, B.F; Bethea, C.G.; Wasserman, E.; Leenders, L. J.; *Chem. Phys.* **1978**, *6*, 5042.

2.   Ishii, T.; Wada, T.; Garito, A.; Sasabe, H.; Yamada, A.; *Mat. Res. Soc. Symp. Proc.* **1990**, *175*, 129.

3.   Cooper, T.; Natarajan, L.; Strasser, R.; Pachter, R.; Crane, R.; *ACS Pol. Prep.* **1990**, *33*, 129.

4.   Pachter, R.; Cooper, T.; Natarajan, L.V.; Crane, R.; Adams, W.W.; *Biopolymers* **1992**, *32*, 1129.

5.   Altman, R.B.; Pachter, R.; Carrara, E.A.; Jardetzky, O.; *QCPE* **1990**, *10* (4), Program 596.

6.   Pachter, R.; Fairchild, S.B.; Lupo, J.A.; Crane, R.L.; Adams, W.W.; *Mat. Res. Soc. Symp. Proc.* **1994**.

7.   Golub, G.; Ortega, G.M.; Scientific Computing: An Introduction with Parallel Computing, Academic Press, NY, **1993**, 66.

8.   Holley, L.H.; Karplus, G.; *Proc. Natl. Acad. Sci. USA.* **1988**, *86*, 152 and references therein.

9.   Fairchild, S.B.; Pachter, R.; Perrin, R.; Crane, R.L.; Adams, W.W.; presented at The American Crystallographic Association, Pittsburgh PA, August, 1992

10. Kabsch, W.; Sanders, C.; *FEBS Letters*, **1983**.

11. Fairchild, S.B.; Pachter, R.; Perrin, R.; *Mathematica Journal*, in press.

12. Altman, R.B.; Jardetzky, O.; *Methods in Enzymology* **1989**, *177*, 218.

13. Gelb, A., Applied Optimal Estimation, MIT Press, **1984**.

14 . Teeter, M.M.; *Proc. Nat. Acad. Sci. USA* **1984**, *81*, 6014.

15. Altman, R.; Pachter, R.; Jardetzky, O.; in 'Protein Structure and Engineering', (O. Jardetzky, Ed.), Plenum Press, New York, **1989**, pp79.

16. Pachter, R.; Altman, R.; Jardetzky, O.; *J. Magn. Reson.* **1990**, *89*, 578.

17. Arrowsmith, C.H.; Pachter, R.; Altman, R.; Iyer, S.B.; Jardetzky, O.; *Biochemistry* **1990**, *29*, 6332; Arrowsmith, C.H.; Pachter, R.; Altman, R.; Jardetzky, O.; *FEBS Eur. J. Biochemistry* **1991**, *202*, 53.

18. Altman, R., C.H. Arrowsmith, Pachter, R.; Jardetzky, O.; in 'Computational Aspects of the Study of Biological Macromolecules by NMR Spectroscopy', (J.C. Hoch, Ed.), Plenum Press, New York **1991**, pp375; Altman, R.; Pachter, R.; Jardetzky, O.; *Applied Magn. Reson.* **1993**, *4*, 441.

19. Pachter, R.; Altman, R.; Czaplicki, J.; Jardetzky, O.; *J. Magn. Reson.* **1991**, *92*, 648.

20. Pachter, R.; Programs: PROPARE, PROCON, PROCOPY, unpublished.

21 . McPhalen, C.A.; James, M.N.G.; *Biochemistry* **1988**, *27*, 6582.

22. Pachter, R.; Lupo, J.A.; Fairchild, S.B.; manuscript in preparation.

23 . Lupo, J.A.; Contributive Research and Development Volume 95, Modeling of NLO Materials Using Parallel Computers, Final Report Task 79, Contract F33615-90-C-5944, Wright Laboratory, Wright-Patterson AFB OH, **1993**, **1994**.

RECEIVED November 15, 1994

INDEXES

Author Index

Affiliation Index

Subject Index

Production: Amie Jackowski & Charlotte McNaughton
Indexing: Deborah H. Steiner
Acquisition: Rhonda Bitterli
Cover design: Amy Hayes

Printed and bound by Maple Press, York, PA

Highlights from ACS Books

Bestsellers from ACS Books

The ACS Style Guide: A Manual for Authors and Editors
Edited by Janet S. Dodd
264 pp; clothbound ISBN 0–8412–0917–0; paperback ISBN 0–8412–0943–X

Understanding Chemical Patents: A Guide for the Inventor
By John T. Maynard and Howard M. Peters
184 pp; clothbound ISBN 0–8412–1997–4; paperback ISBN 0–8412–1998–2

Chemical Activities (student and teacher editions)
By Christie L. Borgford and Lee R. Summerlin
330 pp; spiralbound ISBN 0–8412–1417–4; teacher ed. ISBN 0–8412–1416–6

Chemical Demonstrations: A Sourcebook for Teachers,
Volumes 1 and 2, Second Edition
Volume 1 by Lee R. Summerlin and James L. Ealy, Jr.;
Vol. 1, 198 pp; spiralbound ISBN 0–8412–1481–6;
Volume 2 by Lee R. Summerlin, Christie L. Borgford, and Julie B. Ealy
Vol. 2, 234 pp; spiralbound ISBN 0–8412–1535–9

Chemistry and Crime: From Sherlock Holmes to Today's Courtroom
Edited by Samuel M. Gerber
135 pp; clothbound ISBN 0–8412–0784–4; paperback ISBN 0–8412–0785–2

Writing the Laboratory Notebook
By Howard M. Kanare
145 pp; clothbound ISBN 0–8412–0906–5; paperback ISBN 0–8412–0933–2

Developing a Chemical Hygiene Plan
By Jay A. Young, Warren K. Kingsley, and George H. Wahl, Jr.
paperback ISBN 0–8412–1876–5

Introduction to Microwave Sample Preparation: Theory and Practice
Edited by H. M. Kingston and Lois B. Jassie
263 pp; clothbound ISBN 0–8412–1450–6

Principles of Environmental Sampling
Edited by Lawrence H. Keith
ACS Professional Reference Book; 458 pp;
clothbound ISBN 0–8412–1173–6; paperback ISBN 0–8412–1437–9

Biotechnology and Materials Science: Chemistry for the Future
Edited by Mary L. Good (Jacqueline K. Barton, Associate Editor)
135 pp; clothbound ISBN 0–8412–1472–7; paperback ISBN 0–8412–1473–5

For further information and a free catalog of ACS books, contact:
American Chemical Society
Product Services Office
1155 16th Street, NW, Washington, DC 20036
Telephone 800–227–5558